LONDON MATHEMATICAL SOCIETY LECTURE NOTE SERIES

Managing Editor: Professor I.M.James,
Mathematical Institute, 24-29 St Giles, Oxford

1. General cohomology theory and K-theory, P.HILTON
4. Algebraic topology: a student's guide, J.F.ADAMS
5. Commutative algebra, J.T.KNIGHT
8. Integration and harmonic analysis on compact groups, R.E.EDWARDS
9. Elliptic functions and elliptic curves, P.DU VAL
10. Numerical ranges II, F.F.BONSALL & J.DUNCAN
11. New developments in topology, G.SEGAL (ed.)
12. Symposium on complex analysis, Canterbury, 1973, J.CLUNIE & W.K.HAYMAN (eds.)
13. Combinatorics: Proceedings of the British combinatorial conference 1973, T.P.McDONOUGH & V.C.MAVRON (eds.)
14. Analytic theory of abelian varieties, H.P.F.SWINNERTON-DYER
15. An introduction to topological groups, P.J.HIGGINS
16. Topics in finite groups, T.M.GAGEN
17. Differentiable germs and catastrophes, Th.BROCKER & L.LANDER
18. A geometric approach to homology theory, S.BUONCRISTIANO, C.P.ROURKE & B.J.SANDERSON
20. Sheaf theory, B.R.TENNISON
21. Automatic continuity of linear operators, A.M.SINCLAIR
23. Parallelisms of complete designs, P.J.CAMERON
24. The topology of Stiefel manifolds, I.M.JAMES
25. Lie groups and compact groups, J.F.PRICE
26. Transformation groups: Proceedings of the conference in the University of Newcastle upon Tyne, August 1976, C.KOSNIOWSKI
27. Skew field constructions, P.M.COHN
28. Brownian motion, Hardy spaces and bounded mean oscillation, K.E.PETERSEN
29. Pontryagin duality and the structure of locally compact abelian groups, S.A.MORRIS
30. Interaction models, N.L.BIGGS
31. Continuous crossed products and type III von Neumann algebras, A.VAN DAELE
32. Uniform algebras and Jensen measures, T.W.GAMELIN
33. Permutation groups and combinatorial structures, N.L.BIGGS & A.T.WHITE
34. Representation theory of Lie groups, M.F.ATIYAH et al.
35. Trace ideals and their applications, B.SIMON
36. Homological group theory, C.T.C.WALL (ed.)
37. Partially ordered rings and semi-algebraic geometry, G.W.BRUMFIEL
38. Surveys in combinatorics, B.BOLLOBAS (ed.)
39. Affine sets and affine groups, D.G.NORTHCOTT
40. Introduction to H_p spaces, P.J.KOOSIS
41. Theory and applications of Hopf bifurcation, B.D.HASSARD, N.D.KAZARINOFF & Y-H.WAN
42. Topics in the theory of group presentations, D.L.JOHNSON
43. Graphs, codes and designs, P.J.CAMERON & J.H.VAN LINT
44. Z/2-homotopy theory, M.C.CRABB
45. Recursion theory: its generalisations and applications, F.R.DRAKE & S.S.WAINER (eds.)
46. p-adic analysis: a short course on recent work, N.KOBLITZ
47. Coding the Universe, A. BELLER, R. JENSEN & P. WELCH
48. Low-dimensional topology, R. BROWN & T.L. THICKSTUN (eds.)
49. Finite geometries and designs, P. CAMERON, J.W.P. HIRSCHFELD & D.R. HUGHES (eds.)
50. Commutator Calculus and groups of homotopy classes, H.J. BAUES
51. Synthetic differential geometry, A. KOCK
52. Combinatorics, H.N.V. TEMPERLEY (ed.)
53. Singularity theory, V.I. ARNOLD
54. Markov processes and related problems of analysis, E.B. DYNKIN
55. Ordered permutation groups, A.M.W. GLASS
56. Journees arithmetiques 1980, J.V. ARMITAGE (ed.)
57. Techniques of geometric topology, R.A. FENN
58. Singularities of differentiable functions, J. MARTINET
59. Applicable differential geometry, F.A.E. PIRANI and M. CRAMPIN
60. Integrable systems, S.P. NOVIKOV et al.

London Mathematical Society Lecture Note Series. 51

Synthetic Differential Geometry

ANDERS KOCK
Lecturer in Mathematics
Aarhus University

CAMBRIDGE UNIVERSITY PRESS
CAMBRIDGE
LONDON NEW YORK NEW ROCHELLE
MELBOURNE SYDNEY

Published by the Press Syndicate of the University of Cambridge
The Pitt Building, Trumpington Street, Cambridge CB2 1RP
32 East 57th Street, New York, NY 10022, USA
296 Beaconsfield Parade, Middle Park, Melbourne 3206, Australia

First published in 1981

Printed in Great Britain at the University Press, Cambridge

Library of Congress catalogue card number 81-6099

British Library cataloguing in publication data

Kock Anders
Synthetic Differential Geometry.-(London Mathematical Society
Lecture Note Series 51 ISSN 0076-0552.)
1.Geometry, Differential
I Title. II Series
516'.3'6 QA641

ISBN 0 521 24138 3

CONTENTS

PART II: Categorical Logic

PART III: Models

PREFACE

The aim of the present book is to describe a foundation for synthetic reasoning in differential geometry. We hope that such a foundational treatise will put the reader in a position where he, in his study of differential geometry, can utilize the synthetic method freely and rigorously, and that it will give him notions and language by which such study can be communicated.

That such notions and language is something that till recently seems to have existed only in an inadequate way is borne out by the following statement of Sophus Lie, in the preface to one of his fundamental articles:

> "The reason why I have postponed for so long these investigations, which are basic to my other work in this field, is essentially the following. I found these theories originally by synthetic considerations. But I soon realized that, as expedient [zweckmässig] the synthetic method is for discovery, as difficult it is to give a clear exposition on synthetic investigations, which deal with objects that till now have almost exclusively been considered analytically. After long vacillations, I have decided to use a half synthetic, half analytic form. I hope my work will serve to bring justification to the synthetic method besides the analytical one."
>
> (Allgemeine Theorie der partiellen Differentialgleichungen erster Ordnung, Math. Ann. 9 (1876).)

What is meant by "synthetic" reasoning? Of course, we do not know exactly what Lie meant, but the following is the way we would describe it: It deals with space forms in terms of their structure, i.e. the basic geometric and conceptual constructions that can be performed on them. Roughly these constructions are the morphisms which constitute the base category in terms of which we work; the space forms themselves being objects of it.

This category is cartesian closed, since, whenever we have formed ideas of "spaces" A and B, we can form the idea of B^A, the "space" of all functions from A to B.

The category theoretic viewpoint prevents the identification of A and B with point sets (and hence also prevents the formation of "random" maps from A to B). This is an old tradition in synthetic geometry, where one, for instance, distinguishes between a "line" and the "range of points on it" (cf. e.g. Coxeter [8] p.20).

What categories in the "Bourbakian" universe of mathematics are mathematical models of this intuitively conceived geometric category? The answer is: many of the "gros toposes" considered since the early 60's by Grothendieck and others, - the simplest example being the category of functors from commutative rings to sets. We deal with these topos theoretic examples in Part III of the book. We do not begin with them, but rather with the axiomatic development of differential geometry on a synthetic basis (Part I), as well as a method of interpreting such development in cartesian closed categories (Part II). We chose this ordering because we want to stress that the axioms are intended to reflect some true properties of the geometric and physical reality; the models in Part III are only servants providing consistency proofs and inspiration for new true axioms or theorems. We present in particular some models $\mathcal{E}$ which contain the category of smooth manifolds as a full subcategory in such a way that "analytic" differential geometry for these corresponds exactly to "synthetic" differential geometry in $\mathcal{E}$.

Most of Part I, as well as several of the papers in the bibliography which go deeper into actual geometric matters with synthetic methods, are written in the "naive" style. By this, we mean that all notions, constructions, and proofs involved are presented *as if* the base category were the category of sets; in particular all constructions on the objects involved are described in terms of "*elements*" of them. However, it is necessary and possible to be able to understand this naive writing as referring to cartesian closed categories. It is *necessary* because the basic axioms of synthetic differential geometry have no models in the category of sets (cf. I §1); and it is possible: this is what Part II is about. The method is that we have to understand by an *element* b of an object B a *generalized* element, that is, a map $b: X \to B$, where X is an arbitrary object, called the *stage of definition*, or the *domain of variation* of the element b.

Elements "defined as different stages" have a long tradition in geometry. In fact, a special case of it is when the geometers say: A circle has no *real* points at infinity, but there are two *imaginary* points at infinity such that every circle passes through them. Here $\mathbb{R}$ and $\mathbb{C}$ are two different *stages* of mathematical knowledge, and something that does not yet exist at stage $\mathbb{R}$ may come into existence at the "later" or "deeper" stage $\mathbb{C}$. - More important for the developments here is passage from stage $\mathbb{R}$ to stage $\mathbb{R}[\varepsilon]$, the "ring of dual numbers over $\mathbb{R}$":

$$\mathbb{R}[\varepsilon] = \mathbb{R}[x]/(x^2).$$

It is true, and will be apparent in Part III, that the notion of elements defined at different stages does correspond to this classical notion of elements defined relative to different commutative rings, like $\mathbb{R}$, $\mathbb{C}$, and $\mathbb{R}[\varepsilon]$, cf. the remarks at the end of III §1.

When thinking in terms of physics (of which geometry of space forms is a special case), the reason for the name "domain of variation" (instead of "stage of definition") becomes clear: for a non-atomistic point of view, a body B is not described just in terms of its "atoms" $b \in B$, that is, maps $\mathbf{1} \to B$, but in terms of "particles" of varying size X, or in terms of motions that take place in B and are parametrized by a temporal extent X; both of these situations being described by maps $X \to B$ for suitable domain of variation X.

The exercises at the end of each paragraph are intended to serve as a further source of information, and if one does not want to solve them, one might read them.

Historical remarks and credits concerning the main text are collected at the end of the book. If a specific result is not credited to anybody, it does not necessarily mean that I claim credit for it. Many things developed during discussions between Lawvere, Wraith, myself, Reyes, Joyal, Dubuc, Coste, Coste-Roy, Bkouche, Veit, Penon, and others. Personally, I want to acknowledge also stimulating questions, comments, and encouragement from Dana Scott, J. Bénabou, P. Johnstone, and from my audiences in Milano, Montréal, Paris, Zaragoza, Buffalo, Oxford, and in particular Aarhus. I want also to thank Henry Thomsen for valuable comments to the early drafts of the book.

The Danish Natural Science Research Council has on several occasions made it possible to gather some of the above-mentioned mathematicians for work sessions in Aarhus. This has been vital to the progress of the subject treated here, and I want to express my thanks.

Warm thanks also to the secretaries at Matematisk Institut, Aarhus, for their friendly help, and in particular, to Else Yndgaard for her expert typing of this book.

Finally, I want to thank my family for all their support, and for their patience with me and the above-mentioned friends and colleagues.

PART I
THE SYNTHETIC THEORY

INTRODUCTION

Lawvere has pointed out that "In order to treat mathematically the decisive abstract general relations of physics, it is necessary that the mathematical world picture involve a cartesian closed category E of smooth morphisms between smooth spaces".

This is also true for differential geometry, which is a science that underlies physics. So everything in the present Part I takes place in such cartesian closed category E. The reader may think of E as "the" category of sets, because most constructions and notions which exist in the category of sets exist in such E; there are some exceptions, like use of the "law of excluded middle", cf. Exercise 1.1 below. The text is written as if E were "the" category of sets. This means that to understand this part, one does not have to know anything about cartesian closed categories; rather, one learns it, at least implicitely, because the synthetic method utilizes the cartesian closed structure all the time, even if it is presented in set theoretic disguise (which, as Part II hopefully will bring out, is really no disguise at all).

Generally, investigating geometric and quantitative relationships brings along with it understanding of the logic appropriate for it. So, it also forces E (which represents our understanding of smoothness) to have certain properties, and not to have certain others. In particular, E must have finite inverse limits, and for some of the more refined investigations, to be a topos.

I.1: BASIC STRUCTURE ON THE GEOMETRIC LINE

The geometric line can, as soon as one chooses two distinct points on it, be made into a commutative ring, with the two points as respectively 0 and 1. This is a decisive structure on it, already known and considered by Euclid, who assumes that his reader is able to move line segments around in the plane (which gives addition), and who teaches his reader how he, with ruler and compass, can construct the fourth proportional of three line segments; taking one of these to be $[0,1]$, this defines the product of the two others, and thus the multiplication on the line. We denote the line, with its commutative ring structure* (relative to some fixed choice of 0 and 1) by the letter R.

Also, the geometric plane can, by some of the basic structure (ruler-and-compass-constructions again), be identified with $R \times R = R^2$ (choose a fixed pair of mutually orthogonal copies of the line R in it), and similarly, space with R^3.

Of course, this basic structure does not depend on having the (arithmetically constructed) real numbers $\mathbb{R}$ as a mathematical model for R.

Euclid maintained further that R was not just a commutative ring, but actually a field. This follows because of his assumption: for any two points in the plane, either they are equal, or they determine a unique line.

We cannot agree with Euclid on this point. For that would imply that the set D defined by

$$D := [\![x \in R \mid x^2 = 0]\!] \subseteq R$$

---

* Actually, it is an algebra over the rationals, since the elements $2 = 1+1$, $3 = 1+1+1$, etc., are multiplicatively invertible in R.

consists of 0 alone, and that would immediately contradict our

Axiom 1. For any* $g: D \to R$, there exists a unique $b \in R$ such that

$$\forall d \in D: g(d) = g(0) + d \cdot b.$$

Geometrically, the axiom expresses that the graph of g is a piece of a unique straight line ℓ, namely the one through $(0,g(0))$ and with slope b

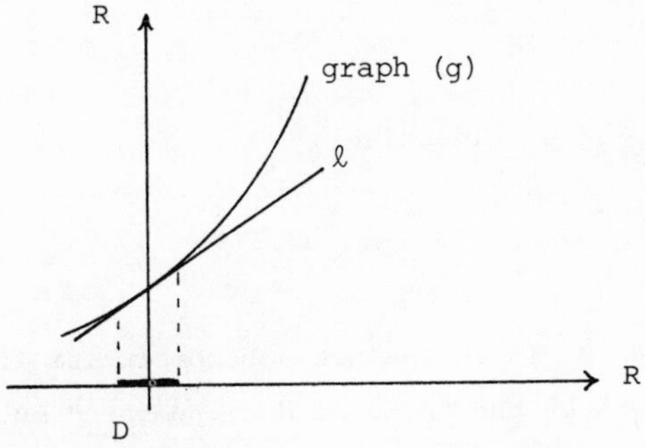

(in the picture, g is defined not just on D, but on some larger set).

Clearly, the notion of slope, which thus is built in, is a decisive abstract general relation for differential calculus. Before we turn to that, let us note the following consequence of the uniqueness assertion in Axiom 1:

$$(\forall d \in D: d \cdot b_1 = d \cdot b_2) \Rightarrow (b_1 = b_2),$$

which we verbalize into the slogan

---

* We really mean: "for any $g \in R^D \ldots$"; this will make a certain difference in the category theoretic interpretation with generalized elements. Similarly for the f in Theorem 2.1 below and several other places.

"universally quantified d's may be cancelled"

("cancelled" here meant in the multiplicative sense).

The axiom may be stated in succinct diagrammatic form in terms of Cartesian Closed Categories. Consider the map α:

$$R \times R \xrightarrow{\alpha} R^D \tag{1.1}$$

given by

$$(a,b) \longmapsto [d \mapsto a + d\cdot b].$$

Then the axiom says

<u>Axiom 1</u>. α is invertible (i.e. bijective).

Let us further note:

<u>Proposition 1.1</u>. The map α is an R-algebra homomorphism if we make $R \times R$ into an R-algebra by the "ring of dual numbers" multiplication

$$(a_1,b_1)\cdot(a_2,b_2) := (a_1\cdot a_2, a_1\cdot b_2 + a_2\cdot b_1) \tag{1.2}$$

<u>Proof</u>. The pointwise product of the maps $D \to R$

$$d \longmapsto a_1 + d\cdot b_1 \qquad\qquad d \longmapsto a_2 + d\cdot b_2$$

is

$$d \longmapsto (a_1 + d\cdot b_1)\cdot(a_2 + d\cdot b_2)$$

$$= a_1\cdot a_2 + d\cdot(a_1\cdot b_2 + a_2\cdot b_1) + d^2\cdot b_1\cdot b_2,$$

but the last term vanishes because $d^2 = 0 \quad \forall d \in D$.

If we let $R[\varepsilon]$ denote $R \times R$, with the ring-of-dual-numbers multiplication, we thus have

Corollary 1.2. Axiom 1 can be expressed: The map α in (1.1) gives an R-algebra isomorphism

$$R[\varepsilon] \xrightarrow[\cong]{} R^D.$$

Assuming Axiom 1, we denote by β and γ, respectively, the two composites

$$\beta = R^D \xrightarrow{\alpha^{-1}} R \times R \xrightarrow{\text{proj}_1} R$$
$$\gamma = R^D \xrightarrow{\alpha^{-1}} R \times R \xrightarrow{\text{proj}_2} R \qquad (1.3)$$

Both are R-linear, by Proposition 1.1; β is just 'evaluation at $0 \in D$' and appears later as the structural map of the tangent bundle of R; γ is more interesting, being the concept of slope itself. It appears later as "principal part formation", (§7), or as the "universal 1-form", or "Maurer-Cartan form" (§18), on $(R,+)$.

EXERCISES AND REMARKS

1.1 (Schanuel). The following construction * is an example of a use of "the law of excluded middle". Define a function $g: D \to R$ by putting

$$* \begin{cases} g(d) = 1 & \text{if } d \neq 0 \\ g(d) = 0 & \text{if } d = 0 \end{cases}$$

If Axiom 1 holds, $d = \{0\}$ is impossible, hence, again by essentially using the law of excluded middle, we may assume $\exists d_o \in D$ with $d_o \neq 0$. By Axiom 1

$$\forall d \in D: g(d) = g(0) + d \cdot b .$$

Substituting d_o for d yields $1 = g(d_o) = 0 + d_o \cdot b$, which, when squared, yields $1 = 0$.

<u>Moral</u>. Axiom 1 is incompatible with the law of excluded middle. Either the one or the other has to leave the scene. In Part I of this book, the law of excluded middle has to leave, being incompatible with the natural synthetic reasoning on smooth geometry to be presented here. In the terms which the logicians use, this means that the logic employed is 'constructive' or 'intuitionistic'. We prefer to think of it just as 'that reasoning which can be carried out in <u>all</u> sufficiently good cartesian closed categories'.

1.2 (Joyal). Assuming Pythagoras' Theorem, it is correct to define the circle around (a,b) with radius c to be

$$[\![(x,y) \in R^2 \mid (x-a)^2 + (y-b)^2 = c^2]\!] .$$

Prove that D is exactly the intersection of the unit circle around $(0,1)$ and the x-axis

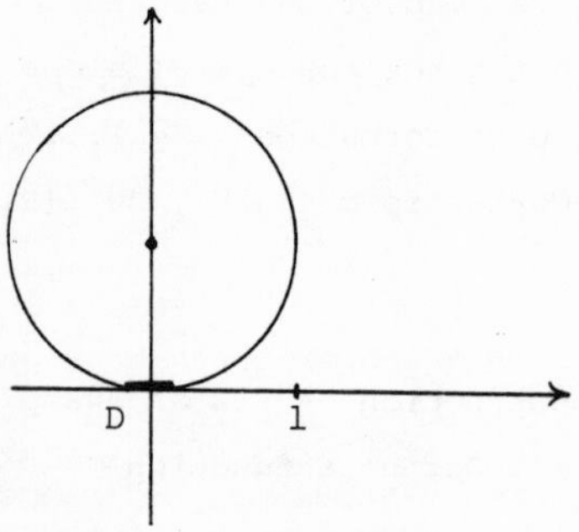

(identifying, as usal, R with the x-axis in R^2).

<u>Remark</u>. This picture of D was proposed by Joyal in 1977. But earlier than that: Hjelmslev [26] experimented in the 1920's with a geometry where, given two points in the plane, there exists <u>at least</u> one line connecting them, but there may exist more than one without the points being identical; this is the case when the points are 'very near' each other. For such geometry, R is not a field, either, and the intersection in the figure above is, like here, not just $\{0\}$. But even earlier than that: Hjelmslev quotes the Old

Greek philosopher, Protagoras, who wanted to refute Euclid by the argument that it is evident that the intersection in the figure contains more than one point.

1.3. If $d \in D$ and $r \in R$, we have $d \cdot r \in R$. If $d_1 \in D$ and $d_2 \in D$, then $d_1 + d_2 \in D$ iff $d_1 \cdot d_2 = 0$ (for the implication '$\Rightarrow$', one must use that 2 is invertible in R).

(In the geometries that have been built based on Hjelmslev's ideas, $d_1^2 = 0 \wedge d_2^2 = 0 \Rightarrow d_1 \cdot d_2 = 0$, but this assumption is incompatible with Axiom 1, see Exercise 4.6 below.)

1.4 (Galuzzi and Meloni; cf.[50] p. 6). Assume $E \subseteq R$ contains 0 and is stable under multiplication by -1. If 2 is invertible in R, and if Axiom 1 holds for E (i.e., when D in Axiom 1 is replaced by E), then $E \subseteq D$.

1.5. If R is any commutative ring, and g is any polynomial (with integral coefficients) in n variables, g gives rise to a polynomial function $R^n \to R$, which may be denoted g_R or just g. For the ring R^X (X an arbitrary object), g_{R^X} gets identified with $(g_R)^X$. To say that a map $\beta: R \to S$ is a ring homomorphism is equivalent to saying that for any polynomial g (in n variables, say)

$$g_S \circ \beta^n = \beta \circ g_R.$$

This is the viewpoint that the algebraic theory consisting of polynomials is the algebraic theory of commutative rings, cf. Appendix A.

In particular, Proposition 1.1 can be expressed: for any polynomial g (in n variables, say), the diagram

$$\begin{array}{ccc} (R[\varepsilon])^n & \xrightarrow{\alpha^n} & (R^D)^n \cong (R^n)^D \\ \downarrow{\scriptstyle g_{R[\varepsilon]}} & & \downarrow{\scriptstyle g_R{}^D} \\ R[\varepsilon] & \xrightarrow[\alpha]{} & R^D \end{array} \tag{1.4}$$

commutes. In III § 4 ff. we shall meet a similar statement, but for arbitrary smooth functions $g: \mathbb{R}^n \to \mathbb{R}$, not just polynomials.

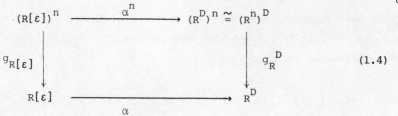

I.2: DIFFERENTIAL CALCULUS

In this §, R is assumed to satisfy Axiom 1; and we assume that $2 \in R$ is invertible.

Let $f: R \to R$ be any function. For fixed $x \in R$, we consider the function $g: D \to R$ given by $g(d) = f(x+d)$. There exists, by Axiom 1, a unique $b \in R$ so that

$$g(d) = g(0) + d \cdot b \quad \forall d \in D, \tag{2.1}$$

or in terms of f

$$f(x+d) = f(x) + d \cdot b \quad \forall d \in D.$$

The b here depends on the x considered. We denote it $f'(x)$, so we have

<u>Theorem 2.1</u> (Taylor's formula). For any $f: R \to R$ and any $x \in R$,

$$f(x+d) = f(x) + d \cdot f'(x) \quad \forall d \in D. \tag{2.2}$$

Formula (2.2) characterizes $f'(x)$. Since we have $f'(x)$ for each $x \in R$, we have in fact defined a new function $f': R \to R$, the derivative of f. The process may be iterated, to define $f'': R \to R$, etc.

If f is not defined on the whole of R, but only on a subset $U \subseteq R$, then we can, by the same procedure, define f' as a function on the set $U' \subseteq U$ given by $U' = [[x \in U \mid x+d \in U \ \forall d \in D]]$. In particular, for $g: D \to R$, we may define $g'(0)$; it is the b

occurring in (2.1). Also, there will in general exist many subsets $U \subseteq R$ with the property that $U' = U$, equivalently, such that

$$x \in U \wedge d \in D \Rightarrow x+d \in U \tag{2.3}$$

For f defined on such a set U, we get $f'\colon U \to R$, $f''\colon U \to R$, etc. In the following Theorem, U and V are subsets of R having the property (2.3).

<u>Theorem 2.2</u>. For any $f,g\colon U \to R$ and any $r \in R$, we have

$$(f+g)' = f' + g' \tag{i}$$

$$(r \cdot f)' = r \cdot f' \tag{ii}$$

$$(f \cdot g)' = f' \cdot g + f \cdot g' \tag{iii}$$

For any $g\colon V \to U$ and $f\colon U \to R$

$$(f \circ g)' = (f' \circ g) \cdot g' \tag{iv}$$

$$id' \equiv 1 \tag{v}$$

$$r' \equiv 0 \tag{vi}$$

(where $id\colon R \to R$ is the identity map and r denotes the constant function with value r).

<u>Proof</u>. All of these are immediate arithmetic calculations based on Taylor's formula. As a sample, we prove the Leibniz rule (iii). For any $x \in U \subseteq R$, we have

$$(f \cdot g)(x+d) = (f \cdot g)(x) + d \cdot (f \cdot g)'(x) \qquad \forall d \in D,$$

by Taylor's formula for $f \cdot g$. On the other hand

$$(f\cdot g)(x+d) = f(x+d)\cdot g(x+d)$$

$$= (f(x) + d\cdot f'(x))\cdot(g(x) + d\cdot g'(x))$$

$$= f(x)\cdot g(x) + d\cdot f'(x)\cdot g(x) + d\cdot f(x)\cdot g'(x);$$

the fourth term $d^2\cdot f'(x)\ g'(x)$ vanishes because $d^2 = 0$. Comparing the two derived expressions, we see

$$d\cdot(f\cdot g)'(x) = d\cdot(f'(x)\cdot g(x) + f(x)\cdot g'(x)) \qquad \forall d \in D.$$

Cancelling the universally quantified d yields the desired

$$(f\cdot g)'(x) = f'(x)\cdot g(x) + f(x)\cdot g'(x).$$

It is not true on basis of Axiom 1 alone that $f' \equiv 0$ implies that f is a constant, or that every f has a primitive g (i.e. $g' \equiv f$ for some g), cf. Part III.

What about Taylor formulae longer than (2.2)? The following is a partial answer for "series" going up to degree-2 terms. It generalizes in an evident way to series going up to degree-n terms. Again, f is a map $U \to R$ with U satisfying (2.3).

Proposition 2.3. For any δ of form d_1+d_2 with d_1 and $d_2 \in D$ we have

$$f(x+\delta) = f(x) + \delta\cdot f'(x) + \frac{\delta^2}{2!} f''(x).$$

Proof.

$$f(x+\delta) = f(x+d_1+d_2) = f(x+d_1) + d_2\cdot f'(x+d_1) \qquad \text{(by (2.2))}$$

$$= f(x) + d_1\cdot f'(x) + d_2\cdot(f'(x) + d_1\cdot f''(x)) \qquad \text{(by (2.2) twice)}$$

$$= f(x) + (d_1+d_2)\cdot f'(x) + d_1\cdot d_2\cdot f''(x).$$

But since $d_1^2 = d_2^2 = 0$, we have $(d_1+d_2)^2 = 2 \cdot d_1 \cdot d_2$. Substituting this, and $\delta = d_1+d_2$, gives the result.

The reasons why the Proposition is to be considered a <u>partial</u> result only, is that we would like to state it for any δ with $\delta^3 = 0$, not just for those of form d_1+d_2 as above. In the models (Part III), $\delta^3 = 0$ does not imply existence of $d_1, d_2 \in D$ with $\delta = d_1+d_2$. In the next §, we strengthen Axiom 1, and after that, the result of Proposition 2.3 will be true for all δ with $\delta^3 = 0$; similarly for still longer Taylor formulae.

<u>EXERCISES</u>

2.1. Assume R is a ring that satisfies the following axiom ("Fermat's Axiom"):

$$\forall f: R \to R \quad \exists! g: R \times R \to R: \tag{2.4}$$
$$\forall x,y \in R \times R: \quad f(x) - f(y) = (x-y) \cdot g(x,y)$$

Define $f': R \to R$ by $f'(x) := g(x,x)$, and prove the results of Theorem 2.2 (this requires a little skill). - The axiom and its investigation is mainly due to Reyes.

Use the idea of Exercise 1.1 to prove that the law of excluded middle is incompatible with Fermat's axiom.

<u>Moral</u>. Fermat's axiom is an alternative synthetic foundation for calculus, which does not use nilpotent elements. The relationship between Axiom 1 and (2.4) is further investigated in §13 (exercises), and models for (2.4) are studied in III §8 and III §9.

I.3: HIGHER TAYLOR FORMULAE (ONE VARIABLE)

In this §, we assume that 2,3,... are invertible in R (i.e. that R is a $\mathbb{Q}$-algebra).

We let $D_k \subseteq R$ denote the set

$$D_k := [\![x \in R \mid x^{k+1} = 0]\!] ,$$

in particular, D_1 is the D considered in §§ 1 and 2. The following is clearly a strengthening of Axiom 1.

Axiom 1'. For any $k = 1,2,\ldots,$ and any $g: D_k \to R$, there exist unique $b_1,\ldots,b_k \in R$ such that

$$\forall d \in D_k: g(d) = g(0) + \sum_{i=1}^{k} d^i \cdot b_i .$$

Assuming this, we can prove

Theorem 3.1 (Taylor's formula). For any $f: R \to R$ and any $x \in R$

$$f(x+\delta) = f(x) + \delta \cdot f'(x) + \ldots + \frac{\delta^k}{k!} f^{(k)}(x) \qquad \forall \delta \in D_k,$$

(again it would suffice for f to be defined only on a suitable subset U around x).

Proof. We give the proof only for $k = 2$ (cf. the exercises below, or [32], for larger k). We have, by Axiom 1', b_1 and b_2 such that, for any $\delta \in D_2$

$$f(x+\delta) = f(x) + \delta \cdot b_1 + \delta^2 \cdot b_2 ; \tag{3.1}$$

specializing to δ's in D_1, we see that $b_1 = f'(x)$. We have, by Proposition 2.3 for any $(d_1,d_2) \in D \times D$

$$f(x+(d_1+d_2)) = f(x) + (d_1+d_2) \cdot f'(x) + (d_1+d_2)^2 \cdot \frac{f''(x)}{2!} \tag{3.2}$$

For $\delta = d_1+d_2$, we therefore have, by comparing (3.1) and (3.2) and using $b_1 = f'(x)$

$$\forall (d_1,d_2) \in D \times D: \ (d_1+d_2)^2 \cdot b_2 = (d_1+d_2)^2 \cdot \frac{f''(x)}{2!}$$

or

$$\forall (d_1,d_2) \in D \times D: \ 2 \cdot d_1 \cdot d_2 \cdot b_2 = 2 \cdot d_1 \cdot d_2 \cdot \frac{f''(x)}{2!} .$$

Cancelling the universally quantified d_1, and then the universally quantified d_2 (and the number 2), we derive

$$b_2 = \frac{f''(x)}{2!} ,$$

q.e.d.

Note that the proof only used the existence part of Axiom 1', not the uniqueness. But for reasons that will become clear in Part II, we prefer to have logical formulae which use only the universal quantifier $\forall$ and the unique-existence quantifier $\exists!$; such formulae have a much simpler semantics, and wider applicability.

EXERCISES

3.1. If $d_1,\ldots,d_k \in D$, then $d_1 + \ldots + d_k \in D_k$. In fact, prove that

$$\begin{aligned} (d_1 + \ldots + d_k)^q &= 0 && \text{if } q \geq k+1 \\ &= q!\sigma_q(d_1,\ldots,d_k) && \text{if } q \leq k, \end{aligned}$$

where $\sigma_q(X_1,\ldots,X_k)$ is the q'th elementary symmetric polynomial in k variables (cf. [77] §29 or [47] V §9). In particular, we have the addition map $\Sigma: D^k \to D_k$ given by

$$(d_1,\ldots,d_k) \longmapsto \Sigma d_i.$$

3.2. If R satisfies Axiom 1' and contains $\mathbb{Q}$ as a subring, prove that if $f: D_k \to R$ satisfies

$$\forall (d_1,\ldots,d_k) \in D^k: f(d_1 + \ldots + d_k) = 0$$

then $f \equiv 0$. (We sometimes phrase this property by saying: "R believes that $\Sigma: D^k \to D_k$ is surjective".)

3.3 (Dubuc and Joyal). Assume R satisfies Axiom 1' and contains $\mathbb{Q}$ as a subring. Then a function $\tau: D^k \to R$ is symmetric (invariant under permutations of the k variables $(d_1,\ldots,d_k)$) iff it factors across the addition map $\Sigma: D^k \to D_k$, that is, iff there exists $t: D_k \to R$ with

$$\forall (d_1,\ldots,d_k) \in D^k: \tau(d_1,\ldots,d_k) = t(\overset{k}{\Sigma} d_i);$$

and such t is unique. (Hint: use the above two exercises, and the fundamental theorem on symmetric polynomials, [77] §29 or [47] V Theorem 11.)

I.4: PARTIAL DERIVATIVES

In this §, we assume Axiom 1. If we formulate this Axiom in the diagrammatic way in terms of function sets:

$$R \times R \xrightarrow{\cong} R^D$$

via the map α, then we also have

$$(R\times R) \times (R\times R) \cong R^D \times R^D \cong (R\times R)^D \cong (R^D)^D \cong R^{D\times D}, \tag{4.1}$$

because of evident rules for calculating with function sets; more generally, we similarly get

$$R^{2^n} \cong R^{D^n}. \tag{4.2}$$

If we want to work out the description of this isomorphism,it is more convenient to use Axiom 1 in the elementwise formulation, and we will get

Proposition 4.1. For any $\tau: D^n \to R$, there exists a unique 2^n-tuple $\{a_H \mid H \subseteq \{1,2,\ldots,n\}\}$ of elements of R such that

$$\forall (d_1,\ldots,d_n) \in D^n: \quad \tau(d_1,\ldots,d_n) = \sum_H a_H \cdot \prod_{j\in H} d_j \ ;$$

in particular, for $n = 2$

$$\forall (d_1,d_2) \in D^2: \quad \tau(d_1,d_2) = a_\emptyset + a_1 \cdot d_1 + a_2 \cdot d_2 + a_{12} \cdot d_1 \cdot d_2 \ .$$

Proof. We do the case $n = 2$, only; the proof evidently generalizes. Given $\tau: D \times D \to R$. For each fixed $d_2 \in D$, we consider $\tau(d_1,d_2)$ as a function of d_1, and have by Axiom 1

$$\forall d_1 \in D: \tau(d_1,d_2) = \bar{a} + \bar{a}_1 \cdot d_1 \tag{4.3}$$

for unique $\bar{a}$ and $\bar{a}_1 \in R$. Now $\bar{a}$ and $\bar{a}_1$ depend on d_2, $\bar{a} = \bar{a}(d_2)$, $\bar{a}_1 = \bar{a}_1(d_2)$. We apply Axiom 1 to each of them to find $a_\emptyset, a_1, a_2$, and a_{12} such that

$$\forall d_2 \in D: \bar{a}(d_2) = a_\emptyset + a_2 \cdot d_2$$

$$\forall d_2 \in D: \bar{a}_1(d_2) = a_1 + a_{12} \cdot d_2.$$

Substituting in (4.3) gives the existence. Putting $d_1 = d_2 = 0$ yields uniqueness of $a_\emptyset$. Then putting $d_2 = 0$ and cancelling the universally quantified d_1 yields uniqueness of a_1; similarly for a_2. Then uniqueness of a_{12} follows by cancelling the universally quantified d_1 and then the universally quantified d_2.

We may introduce partial derivatives in the expected way. Let $f: R^n \to R$ be any function. For fixed $\underline{r} = (r_1,\ldots,r_n) \in R^n$, we consider the function $g: D \to R$ given by

$$g(d) := f(r_1+d, r_2,\ldots,r_n) \tag{4.4}$$

By Axiom 1, there exists a unique $b \in R$ so that $g(d) = g(0) + d \cdot b$. We denote this b by $\frac{\partial f}{\partial x_1}(r_1,\ldots,r_n)$, so that we have, by substituting in (4.4)

$$\forall d \in D: f(r_1+d, r_2,\ldots,r_n) = f(r_1,\ldots,r_n) + d \cdot \frac{\partial f}{\partial x_1}(r_1,\ldots,r_n)$$

which thus characterizes a new function $\frac{\partial f}{\partial x_1}: R^n \to R$. Similarly, we define $\frac{\partial f}{\partial x_2}, \ldots, \frac{\partial f}{\partial x_n}$. The process may be iterated, so that we may form for instance

$$\frac{\partial}{\partial x_2}\left(\frac{\partial f}{\partial x_1}\right), \quad \text{denoted } \frac{\partial^2 f}{\partial x_2 \partial x_1} \quad .$$

If f is not defined on the whole of R^n, but only on a subset $U \subseteq R^n$, then we can define $\frac{\partial f}{\partial x_1}$ on the subset of U consisting of those $(r_1,\ldots,r_n)$ for which, for all $d \in D$, $(r_1+d, r_2,\ldots,r_n) \in U$. Similarly for $\frac{\partial f}{\partial x_j}$. In particular, if τ is defined on $D \times D \subseteq R \times R$, then $\frac{\partial \tau}{\partial x_1}$ is defined on $\{0\} \times D$, and is in fact the function $\bar{a}_1$ considered in the proof of Proposition 4.1; similarly $\frac{\partial \tau}{\partial x_2}$ is defined on $D \times \{0\}$, so both $\frac{\partial^2 \tau}{\partial x_2 \partial x_1}$ and $\frac{\partial^2 \tau}{\partial x_1 \partial x_2}$ are defined at $(0,0)$; and

$$\frac{\partial^2 \tau}{\partial x_2 \partial x_1}(0,0) = \frac{\partial \bar{a}_1}{\partial x_2} = a_{12}.$$

But in Proposition 4.1, the variables occur on equal footing, so that we may similarly conclude

$$\frac{\partial^2 \tau}{\partial x_1 \partial x_2}(0,0) = a_{12}.$$

The following is then an immediate Corollary

<u>Proposition 4.2</u>. For any function $f: U \to R$, where $U \subseteq R^n$,

$$\frac{\partial^2 f}{\partial x_i \partial x_j} = \frac{\partial^2 f}{\partial x_j \partial x_i}$$

in those points of U where both are defined.

There is a sense in which partial derivatives may be seen as a special case of ordinary derivatives, namely by passage to "the category of objects over a given object", cf. II §6, and [32].

EXERCISES

4.1. Prove that for any function $f: R^2 \to R$, we have

$$f(r_1+d_1,r_2+d_2) = f(r_1,r_2) + d_1\cdot\frac{\partial f}{\partial x_1}(r_1,r_2) + d_2\cdot\frac{\partial f}{\partial x_2}(r_1,r_2) + d_1\cdot d_2\cdot\frac{\partial^2 f}{\partial x_1\partial x_2}(r_1,r_2)$$

for any $(d_1,d_2) \in D\times D$.

4.2. Use Proposition 4.1 (for $n = 2$) to prove that the following "Property W" holds for $M = R$:

For any $\tau: D\times D \to M$ with $\tau(d,0) = \tau(0,d) = \tau(0,0)$ $\forall d \in D$, there exists a unique $t: D \to M$ with

$$\forall (d_1,d_2) \in D\times D: \tau(d_1,d_2) = t(d_1\cdot d_2).$$

Prove also that Property W holds for $M = R^n$ (for any n).

4.3. If all $d \in D$ were of form $d_1\cdot d_2$ for some $(d_1,d_2) \in D\times D$, then clearly if M satisfies Property W, then so does any subset $N \subseteq M$. However, we do not want to assume that (it is false in the models). Prove that we always have the following weaker result: if M and P satisfy W, and $f,g: M \to P$ are two maps, then the set N (the equalizer of f and g),

$$N := [\![m \in M \mid f(m) = g(m)]\!]$$

satisfies W. For a more complete result, see Exercise 6.6.

4.4. Assume R contains $\mathbb{Q}$. Consider, in analogy with the Property W of Exercise 4.2, the following "Symmetric-Functions-Property" for M:

For any $\tau: D^n \to M$ with τ symmetric,
there exists a unique $t: D_n \to M$, with (4.5)
$\forall (d_1,\ldots,d_n) \in D^n: \tau(d_1,\ldots,d_n) = t(d_1 + \ldots + d_n)$.

Prove, assuming Axiom 1', that $R = M$ has this property (this is just a reformulation of Exercise 3.3). Also, prove that this property has similar stability properties as those discussed for Property W in Exercise 4.3. For a more complete result, see Exercise 6.6.

4.5. Prove that any function $\tau: D^n \to R$ with

$$\tau(0,d_2,\ldots,d_n) = \tau(d_1,0,d_3,\ldots,d_n) = \ldots$$

$$= \tau(d_1,\ldots,d_{n-1},0) \qquad \forall (d_1,\ldots,d_n) \in D^n$$

is of form

$$\tau(d_1,\ldots,d_n) = a + (\prod_{i=1}^{n} d_i)\, b$$

for unique $a,b \in R$. We phrase this: "Property W_n holds for $M = R$".

4.6. Prove that the formula

$$\forall (d_1,d_2) \in D \times D: d_1 \cdot d_2 = 0$$

is incompatible with Axiom 1 (Hint: cancel the universally quantified d_1 to conclude $\forall d_2 \in D: d_2 = 0$.)

4.7 (Wraith). Assume that 2 is invertible in R; prove that the sentence

$$\forall (x,y) \in R \times R \;\; x^2 + y^2 = 0 \Rightarrow x^2 = 0 \qquad (4.6)$$

is incompatible with Axiom 1. (Hint: for $(d_1,d_2) \in D \times D$, consider $(d_1 + d_2)^2 + (d_1 - d_2)^2$ as the $x^2 + y^2$ in (4.6); then utilize Exercise 4.6).

I.5: HIGHER TAYLOR FORMULAE IN SEVERAL VARIABLES. TAYLOR SERIES.

In this §, we assume that R is a $\mathbb{Q}$-algebra and satisfies Axiom 1'. We remind the reader about standard conventions concerning multi-indices: an n-index is an n-tuple $\alpha = (\alpha_1, \ldots, \alpha_n)$ of non-negative integers. We write $\alpha!$ for $\alpha_1! \cdot \ldots \cdot \alpha_n!$, $|\alpha|$ for $\Sigma\alpha_j$, and, whenever $\underline{x} = (x_1, \ldots, x_n)$ is an n-tuple of elements in a ring, $\underline{x}^\alpha$ denotes $x_1^{\alpha_1} \cdot \ldots \cdot x_n^{\alpha_n}$. Also

$$\frac{\partial^{|\alpha|} f}{\partial x^\alpha} \text{ denotes } \frac{\partial^{|\alpha|} f}{\partial x_1^{\alpha_1} \cdots \partial x_n^{\alpha_n}}$$

Finally, we say $\alpha \leq \beta$ if $\alpha_i \leq \beta_i$ for $i = 1, \ldots, n$.

The following two facts are then proved in analogy with the corresponding results (Proposition 4.1 and Exercise 4.1) in §4. Let $k = (k_1, \ldots, k_n)$ be a multi-index.

<u>Proposition 5.1</u>. For any $\tau: D_{k_1} \times \ldots \times D_{k_n} \to R$, there exists a unique polynomial with coefficients from R of form

$$\varphi(X_1, \ldots, X_n) = \sum_{\alpha \leq k} a_\alpha \cdot X^\alpha$$

such that $\forall (d_1, \ldots, d_n) \in D_{k_1} \times \ldots \times D_{k_n}$

$$\tau(d_1, \ldots, d_n) = \varphi(d_1, \ldots, d_n).$$

<u>Theorem 5.2</u> (Taylor's formula in several variables). Let $f: U \to R$ where $U \subseteq R^n$. For every $\underline{r} \in U$ such that $\underline{r} + \underline{d} \in U$ for all $\underline{d} \in D_{k_1} \times \ldots \times D_{k_n}$, we have

$$f(\underline{r}+\underline{d}) = \sum_{\alpha \leq k} \frac{\underline{d}^{\alpha}}{\alpha!} \cdot \frac{\partial^{|\alpha|} f}{\partial x^{\alpha}} (\underline{r}) \quad \forall \underline{d} \in D_{k_1} \times \ldots \times D_{k_n} \tag{5.1}$$

We omit the proofs. Note that (5.1) remains valid even if we include some terms into the sum whose multi-index α does not satisfy $\alpha \leq k$. For, in such terms $\underline{d}^{\alpha}$ is automatically zero.

We let $D_\infty \subseteq R$ denote $\bigcup D_k$. (For this naively conceived union to make sense in $\mathcal{E}$, we need that $\mathcal{E}$ has unions of subobjects, and that such have good exactness properties. This will be the case if $\mathcal{E}$ is a topos.) So we have

$$D_\infty = [\![x \in R \mid x \text{ is nilpotent}]\!] .$$

The set $D_\infty^n \subseteq R^n$ is going to play a role in many of the following considerations, as the 'monad' or '∞-monad' around $\underline{0} \in R^n$. For functions defined on it, we have

Theorem 5.3 (Taylor's series). Let $f: D_\infty^n \to R$. Then there exists a unique formal power series $\Phi(X_1,\ldots,X_n)$ in n variables, and with coefficients from R, such that

$$f(\underline{d}) = \Phi(\underline{d}) \quad \forall \underline{d} = (d_1,\ldots,d_n) \in D_\infty^n .$$

Note that the right hand side makes sense because each coordinate of $\underline{d}$ is nilpotent, so there are only finitely many non-zero terms in $\Phi(\underline{d})$.

Proof. We note first that

$$D_\infty^n = (\bigcup_k D_k)^n = \bigcup_k (D_k^n)$$

We let the coefficient of X^{α} in Φ be

$$\frac{1}{\alpha!} \frac{\partial^{|\alpha|} f}{\partial x_{\alpha}} (\underline{0}) .$$

If $\underline{d} \in D_\infty^n$, we have $\underline{d} \in D_k^n$ for some k, and so Theorem 5.2 tells us that $f(\underline{d}) = \Phi(\underline{d})$. To prove uniqueness, if Φ is a series which is zero on D_∞^n, it is zero on D_k^n for each k. But its restriction to D_k^n is given by a polynomial obtained by truncating the series suitably. From Proposition 5.1, we conclude that this polynomial is zero. We conclude that Φ is the zero series (i.e. all coefficients are zero).

EXERCISES

5.1. Prove that $D_\infty \subseteq R$ is an ideal (in the usual sense of ring theory). Prove that $D_\infty^n \subseteq R^n$ is a submodule.

5.2. Prove that a map $t: D_\infty \to R$ with $t(0) = 0$ maps D_k into D_k, for any k.

5.3. Let V be an R-module. We say that V satisfies the vector form of Axiom 1' if for any $k = 1,2,\ldots$ and any $g: D_k \to V$, there exist unique $\underline{b}_1,\ldots,\underline{b}_k \in V$ so that

$$\forall d \in D_k:\ g(d) = g(0) + \sum_{i=1}^{k} d^i \cdot \underline{b}_i \ .$$

Prove that any R-module of form R^n satisfies this, and that if V does, then so does V^X, for any object X.

The latter fact becomes in particular evident if we write Axiom 1' (for $k = 1$, i.e. Axiom 1) in the form

$$V \times V \xrightarrow{\ \cong\ } V^D$$

via α, compare (1.1), because

$$(V \times V)^X \cong V^X \times V^X$$

and

$$(V^D)^X \cong (V^X)^D$$

are general truths about functions sets, i.e. about cartesian closed categories.

5.4. Let V be an R-module which satisfies the vector form of Axiom 1'. For $f: R \to V$, define $f': R \to V$ so that for any $x \in R$, we have

$$f(x+d) = f(x) + d \cdot f'(x) \qquad \forall d \in D.$$

Similarly, for $f: R^n \to V$, define $\frac{\partial f}{\partial x_i} : R^n \to V$ $(i = 1, \ldots, n)$, and formulate and prove analogues of Theorem 3.1 and Theorem 5.2.

I.6: SOME IMPORTANT INFINITESIMAL OBJECTS

Till now, we have met

$$D = D_1 = [\![x \in R \mid x^2 = 0]\!]$$

and more generally

$$D_k = [\![x \in R \mid x^{k+1} = 0]\!] ,$$

as well as cartesian products of these, like $D_{k_1} \times \ldots \times D_{k_n} \subseteq R^n$. We describe here some further important "infinitesimal objects". First, some that are going to be our "standard 1-monads", and represent the notion of "1-jet":

$$D(2) = [\![(x_1,x_2) \in R^2 \mid x_1^2 = x_2^2 = x_1 \cdot x_2 = 0]\!],$$

more generally

$$D(n) = [\![(x_1,\ldots,x_n) \in R^n \mid x_i \cdot x_j = 0 \quad \forall i,j = 1,\ldots,n]\!].$$

We have $D(2) \subseteq D \times D$, and $D(n) \subseteq D^n \subseteq R^n$. Note $D(1) = D$. Next, the following are going to be our "standard k-monads", and represent the notion of "k-jet":

$$D_k(n) = [\![(x_1,\ldots,x_n) \in R^n \mid \text{the product of any } k{+}1 \text{ of the } x_i\text{'s is zero}]\!] .$$

Clearly

$$D_k(n) \subseteq D_\ell(n) \qquad \text{for} \quad k \leq \ell.$$

Note $D(n) = D_1(n)$. By convention, $D_o(n) = \{\underline{0}\} \subseteq R^n$.

We note

$$D_k(n) \subseteq (D_k)^n$$

and

$$(D_k)^n \subseteq D_{n\cdot k}(n)$$

from which we conclude

$$D_\infty^n = \bigcup_{k=1} D_k(n) \tag{6.1}$$

We list some canonical maps between some of these objects. Besides the projection maps from a product to its factors, and the inclusion maps $D_k(n) \subseteq D_{\ell(n)}$ for $k \leq \ell$, we have

$$\text{incl}_i\colon D \longrightarrow D(n) \qquad (i = 1,\ldots,n) \tag{6.2}$$

given by

$$d \longmapsto (0,\ldots,d,\ldots,0) \qquad (d \text{ in the i'th place})$$

as well as

$$\Delta\colon D \longrightarrow D(n) \tag{6.3}$$

given by

$$d \longmapsto (d,d,\ldots,d).$$

We also have maps like

$$\text{incl}_{12}\colon D(2) \longrightarrow D(3) \tag{6.4}$$

given by $(d,\delta) \longmapsto (d,\delta,0)$, and

$$\Delta\times 1\colon D(2) \longrightarrow D(3) \tag{6.5}$$

given by $(d,\delta) \longmapsto (d,d,\delta)$. We use these maps in §7.

We have already (Exercise 3.1) considered the addition map $\Sigma\colon D^n \to D_n$. It restricts to a map

$$\Sigma\colon D(n) \longrightarrow D,$$

since $(d_1 + \ldots + d_n)^2 = 0$ if the product of any two of the d_i's is zero. More generally, the $D_k(n)$'s have the following good property, not shared by the $(D_k)^n$'s:

Proposition 6.1. Let $\varphi = (\varphi_1,\ldots,\varphi_m)$ be an m-tuple of polynomials in n variables, with coefficients from R and with 0 constant term. Then the map $\varphi\colon R^n \to R^m$ defined by the m-tuple has the property

$$\varphi(D_k(n)) \subseteq D_k(m).$$

Proof. Let $\underline{d} = (d_1,\ldots,d_n) \in D_k(n)$. Each term in each $\varphi_i(d_1,\ldots,d_n)$ contains at least one factor d_j for some $j = 1,\ldots,n$, since φ_i has zero constant term. Any product

$$\varphi_{i_1}(\underline{d})\cdot\ldots\cdot\varphi_{i_{k+1}}(\underline{d}),$$

if we rewrite it by the distributive law, is thus a sum of terms each with $k+1$ factors, each of which contains at least one d_j.

Axiom 1". For any $k = 1,2,\ldots,$ and any $n = 1,2,\ldots,$ any map $D_k(n) \to R$ is uniquely given by a polynomial (with coefficients from R) in n variables and of total degree $\leq k$.

Even with this Axiom, there are still "infinitesimal" objects $\tilde{D}$, where we do not have any conclusion about maps $\tilde{D} \to R$, like for example the object

$$D_c = [\![(x,y) \in R^2 \mid x\cdot y = 0 \wedge x^2 = y^2]\!] \subseteq D_2(2). \tag{6.6}$$

Instead, we give in §16 a uniform conceptual "Axiom 1^W", that implies Axiom 1" as well as most other desirable conclusions about maps from infinitesimal objects to R.

The Proposition 6.1 has the following immediate

Corollary 6.2. Assume R satisfies Axiom 1". Then every map $\varphi: D_k(n) \to R^m$ with $\varphi(\underline{0}) = \underline{0}$ factors through $D_k(m)$.

We shall prove that Axiom 1" implies that the object $M = R$ is infinitesimally linear in the following sense:

Definition 6.3. An object M is called infinitesimally linear if for each $n = 2,3,\ldots,$ and each n-tuple of maps

$$t_i: D \to M \quad \text{with} \quad t_1(0) = \ldots = t_n(0),$$

there exists a unique $\ell: D(n) \to M$ with $\ell \circ \text{incl}_i = t_i$ $(i = 1,\ldots,n)$.

Proposition 6.4. Axiom 1" implies that R is infinitesimally linear.

Proof. Given $t_i: D \to R$ $(i = 1,\ldots,n)$ with $t_i(0) = a \in R$ $\forall i$. By Axiom 1, t_i is of form

$$t_i(d) = a + d\cdot b_i \qquad \forall d \in D.$$

Construct $\ell: D(n) \to R$ by

$$\ell(d_1,\ldots,d_n) = a + \Sigma d_i \cdot b_i .$$

Then clearly $\ell \circ \mathrm{incl}_i = t_i$. This proves existence. To prove uniqueness, let $\tilde{\ell}: D(n) \to R$ be arbitrary with $\tilde{\ell}(\underline{0}) = a$. By Axiom 1" (for $k = 1$), $\tilde{\ell}$ is the restriction of a unique polynomial map of degree ≤ 1, so

$$\tilde{\ell}(d_1,\ldots,d_n) = a + \Sigma d_i \cdot \tilde{b}_i \qquad \forall (d_1,\ldots,d_n) \in D(n)$$

for some unique $\tilde{b}_1,\ldots,\tilde{b}_n \in R$. If we assume $\ell \circ \mathrm{incl}_i = \tilde{\ell} \circ \mathrm{incl}_i$ $\forall i$, we see

$$a + d \cdot b_i = a + d \cdot \tilde{b}_i, \qquad \forall d \in D,$$

whence, by cancelling the universally quantified d, $b_i = \tilde{b}_i$. We conclude $\ell = \tilde{\ell}$.

One would hardly say that a conceptual framework for synthetic differential geometry were complete if it did not have some notion of "neighbour-" relation for the elements of sufficiently good objects M; better, for each natural number k, a notion of "k-neighbour" relation $x \sim_k y$ for the elements of M. It will be defined below, for certain M.

A typical phrase occurring in Lie's writings, were he explicitely says that he is using synthetic reasoning, is "these two families of curves have two ... neighbouring curves p_1 and $\bar{p}_1$ in common", ([54],p.49). "Neighbour" means "1-neighbour", since the authors of the 19th century tradition would talk about "two consecutive neighbours" for what in our attempt would be dealt with in terms of "a 2-neighbour". These two notions are closedly related, because of the observation (Exercise 3.2) that "R believes $\Sigma: D \times D \to D_2$ is surjective".

The neighbour relations $\sim_k$ in synthetic differential geometry are not those considered in non-standard analysis [73]: their neighbour relation is transitive, and is not stratified into "1-neighbour", "2-neighbour", etc., a stratification which is closely tied to "degree-1-segment", "degree-2-segment" of Taylor series.

On the coordinate spaces R^n, we may introduce, for each natural number k, the k-neighbour relation, denoted $\sim_k$, by

$$\underline{x} \sim_k \underline{y} \iff (\underline{x}-\underline{y}) \in D_k(n).$$

It is a reflexive and symmetric relation, and it is readily proved that

$$\underline{x} \sim_k \underline{y} \wedge \underline{y} \sim_\ell \underline{z} \Rightarrow \underline{x} \sim_{k+\ell} \underline{z}.$$

We write $\mathcal{M}_k(\underline{x})$ ("The k-monad around $\underline{x}$") for $[\![\underline{y} \mid \underline{x} \sim_k \underline{y}]\!]$. Thus $\mathcal{M}_k(\underline{x})$ is the fibre over $\underline{x}$ of

$$\begin{array}{c} (R^n)_{(k)} \\ \downarrow \\ R^n \end{array} \tag{6.7}$$

where $(R^n)_{(k)} \subseteq R^n \times R^n$ is the object $[\![(x,y) \mid \underline{x} \sim_k \underline{y}]\!]$ and where the indicated map is projection onto the first factor.

Similarly, we define

$$\underline{x} \sim_\infty \underline{y} \iff (\underline{x}-\underline{y}) \in D^n_\infty = \bigcup_k D_k(n),$$

and $\mathcal{M}_\infty(\underline{x}) = [\![\underline{y} \mid \underline{x} \sim_\infty \underline{y}]\!]$, "the ∞-monad around $\underline{x}$". The relation $\sim_\infty$ is actually an equivalence relation.

From Corollary 6.2, we immediately deduce

Corollary 6.5. Any map $f\colon \mathcal{M}_k(\underline{x}) \to R^m$ factors through $\mathcal{M}_k(f(\underline{x})) \subseteq R^m$ (this also holds for $k = \infty$). Equivalently, $\underline{x} \sim_k \underline{y}$ implies $f(\underline{x}) \sim_k f(\underline{y})$.

For the category of objects M of form R^m, more generally, for the category of formal manifolds considered in §17 below, where we also construct relations $\sim_k$, the conclusion of the Corollary may be formulated: for any $f: M \to N$, the map $f \times f: M \times M \to N \times N$ restricts to a map $M_{(k)} \to N_{(k)}$.

EXERCISES

6.1. Show that the map $R^2 \to R^2$ given by $(x_1, x_2) \longmapsto (x_1, x_1 \cdot x_2)$ restricts to a map $D \times D \to D(2)$. Compose this with the addition map $\Sigma: D(2) \to D$ to obtain a non-trivial map $\lambda: D \times D \to D$. (This map induces the "Liouville vector field", cf. Exercise 8.6.)

6.2. Show that if 2 is invertible in R, the counter image of $D \subset D_2$ under the addition map $\Sigma: D \times D \to D_2$ is precisely $D(2)$.

6.3. Show that $D_k(n) \times D_\ell(m) \subseteq D_{k+\ell}(n+m)$. Show also that

$$\underline{a} \in D_k(n) \wedge \underline{b} \in D_\ell(n) \Rightarrow \underline{a}+\underline{b} \in D_{k+\ell}(n).$$

6.4. Let

$$\tilde{D}(2,n) = [\![((d_1,\ldots,d_n),(\delta_1,\ldots,\delta_n)) \in R^n \times R^n \mid$$

$$d_i \cdot \delta_j + d_j \cdot \delta_i = 0 \wedge d_i \cdot d_j = 0$$

$$\wedge\ \delta_i \cdot \delta_j = 0 \quad \forall i,j = 1,\ldots,n]\!] .$$

Prove that any symmetric bilinear map $R^n \times R^n \to R$ vanishes on $\tilde{D}(2,n)$, provided the number 2 is invertible in R.

The geometric significance of $\tilde{D}(2,n)$, and some analogous $\tilde{D}(h,n)$ for larger h, is studied in §16, notably Proposition 16.5.

6.5. Prove that if M_1 and M_2 are infinitesimally linear, then so is $M_1 \times M_2$, and for any two maps $f,g\colon M_1 \to M_2$, the equalizer $[\![m \in M_1 \mid f(m) = g(m)]\!]$ is infinitesimally linear.

In categorical terms: the class of infinitesimally linear objects is closed under formation of finite inverse limits in $\mathcal{E}$.

Also, if M is infinitesimally linear, then so is M^X, for any object X.

The categorically minded reader may see the latter at a glance by utilizing:

i) M is infinitesimally linear iff for each n

$$M^{D(n)} \cong M^D \times_M \cdots \times_M M^D \qquad \text{(n fold pull-back)}.$$

ii) $(-)^X$ preserves pull-backs.

iii) $(M^{D(n)})^X \cong (M^X)^{D(n)}$

6.6. Express in the style of the last part of Exercise 6.5 (i.e. in terms of finite inverse limit diagrams) Property W on M (Exercise 4.2), as well as the Symmetric Functions Property (Exercise 4.4), and deduce that the class of objects satisfying Property W, respectively The Symmetric Functions Property, is stable under finite inverse limits and $(-)^X$ (for any X), (cf. [71]).

6.7. Assume that R is infinitesimally linear, satisfies Axiom 1" and contains $\mathbb{Q}$ as a subring. Prove that $(D_k(n))^m$ and D_∞^n are infinitesimally linear, have Property W and the Symmetric Functions Property.

I.7: TANGENT VECTORS AND THE TANGENT BUNDLE

In this §, we consider, besides the line R, some unspecified object M (to be thought of as a "smooth space", since our base category $\mathcal{E}$ is the category of such, even though we talk about $\mathcal{E}$ as if its objects were sets). For instance, M might be R, or R^m, or some 'affine scheme' like the circle in Exercise 1.1, or $D_k(n)$; or something glued together from affine pieces, - the projective line over R, say. It could also be some big function space like R^R (= set of all maps from R to itself), or R^{D_∞}.

There will be ample justification for the following

Definition 7.1. A tangent vector to M, with base point $x \in M$ (or attached at $x \in M$) is a map $t: D \to M$ with $t(0) = x$.

This definition is related to one of the classical ones, where a tangent vector at $x \in M$ (M a manifold) is an equivalence class of "short paths" $t: (-\varepsilon,\varepsilon) \to M$ with $t(0) = x$. Each representative $t: (-\varepsilon,\varepsilon) \to M$ contains redundant information, whereas our D is so small that a $t: D \to M$ gives a tangent vector with no redundant information; thus, here, tangent vectors are infinitesimal paths, of "length" D.

This is a special case of the feature of synthetic differential geometry that the jet notion becomes representable.

We consider the set M^D of all tangent vectors to M. It comes equipped with a map $\pi: M^D \to M$, namely $\pi(t) = t(0)$. Thus π associates to any tangent vector its base point; M^D together with π is called the tangent bundle of M. The fibre over $x \in M$, i.e. the set of tangent vectors with x as base point, is called the tangent space to M at x, and denoted $(M^D)_x$. Sometimes we write TM, respectively T_xM, for M^D and $(M^D)_x$.

The construction M^D (like any exponent-formation in a cartesian closed category) is functorial in M. The elementary description is also evident; given $f: M \to N$, we get $f^D: M^D \to N^D$ described as follows

$$f^D(t) = f \circ t: D \to M \to N,$$

equivalently, f^D is described by

$$t \longmapsto [d \longmapsto f(t(d))] .$$

Also, $\pi: M^D \to M$ is natural in M. Note that if t has base point x, $f \circ t$ has base point $f(x)$.

To justify the name tangent vector, one should exhibit a "vector space" (R-module) structure on each tangent space $(M^D)_x$. This we can do when M is infinitesimally linear. In any case, we have an action of the multiplicative semigroup $(R,\cdot)$ on each $(M^D)_x = T_xM$, namely, for $r \in R$ and $t: D \to M$ with $t(0) = x$, define $r\cdot t$ by putting

$$(r\cdot t)(d) := t(r\cdot d),$$

("changing the speed of the infinitesimal curve t by the factor r").

Now let us assume M infinitesimally linear; to define an addition on T_xM, we proceed as follows. We remind the reader of the maps $incl_i: D \to D(2)$, $\Delta: D \to D(2)$ ((6.2),(6.3)). If $t_1,t_2: D \to M$ are tangent vectors to M with base point x, we may, by infinitesimal linearity, find a unique $\ell: D(2) \to M$ with

$$\ell \circ incl_i = t_i \qquad i = 1,2 . \tag{7.1}$$

We define t_1+t_2 to be the composite

$$D \xrightarrow{\Delta} D(2) \xrightarrow{\ell} M$$

("diagonalizing ℓ"); in other words

$$\forall d \in D: (t_1+t_2)(d) = \ell(d,d)$$

where $\ell: D(2) \to M$ is the unique map with

$$\forall d \in D: \ell(d,0) = t_1(d) \wedge \ell(0,d) = t_2(d).$$

Proposition 7.2. Let M be infinitesimally linear. With the addition and multiplication-by-scalars defined above, each T_xM becomes an R-module. Also, if $f: M \to N$ is a map between infinitesimally linear objects, $f^D: M^D \to N^D$ restricts to an R-linear map $T_xM \to T_{f(x)}N$.

Proof. Let us prove that the addition described is associative. So let $t_1, t_2, t_3: D \to M$ be three tangent vectors at $x \in M$. By infinitesimal linearity of M, there exists a unique

$$\ell: D(3) \to M$$

with

$$\ell \circ incl_i = t_i \qquad (i = 1,2,3). \tag{7.2}$$

We claim that $(t_1+t_2)+t_3$ and $t_1+(t_2+t_2)$ are both equal to

$$D \xrightarrow{\Delta} D(3) \xrightarrow{\ell} M \tag{7.3}$$

For with notation as in (6.4) and (6.5)

$$(\ell \circ incl_{12}) \circ incl_1 = \ell \circ incl_1 = t_1$$

$$(\ell \circ incl_{12}) \circ incl_2 = \ell \circ incl_2 = t_2$$

so that

$$(\ell \circ \mathrm{incl}_{12}) \circ \Delta = t_1 + t_2 .$$

Also

$$(\ell \circ (\Delta \times 1)) \circ \mathrm{incl}_1 = \ell \circ \mathrm{incl}_{12} \circ \Delta = t_1 + t_2$$

and

$$(\ell \circ (\Delta \times 1)) \circ \mathrm{incl}_2 = \ell \circ \mathrm{incl}_3 \quad = t_3$$

so that

$$(\ell \circ (\Delta \times 1)) \circ \Delta = (t_1 + t_2) + t_3 .$$

But the left hand side here is clearly equal to (7.3). Similarly for $t_1 + (t_2 + t_3)$. This proves associativity of $+$. We leave to the reader to verify commutativity of $+$, and the distributive laws for multiplication by scalars $\in R$. Note that the zero tangent vector at x is given by $t(d) = x \quad \forall d \in D$.

Also, we leave to the reader to prove the assertion about R-linearity of $T_x M \to T_{f(x)} N$.

Let V be an R-module which satisfies the (vector form of) Axiom 1, that is, for every $t: D \to V$, there exists a unique $\underline{b} \in V$ so that

$$\forall d \in D: t(d) = t(0) + d \cdot \underline{b}$$

(cf. Exercise 5.3 and 5.4); R^k is an example. We call $\underline{b} \in V$ the <u>principal part</u> of the tangent vector t.

In the following, Proposition V is such an R-module, which furthermore is assumed to be infinitesimally linear.

Proposition 7.3. Let t_1, t_2 be tangent vectors to V with same base point $\underline{a} \in V$, and with principal parts $\underline{b}_1$ and $\underline{b}_2$, respectively. Then $t_1 + t_2$ has principal part $\underline{b}_1 + \underline{b}_2$. Also, for any $r \in R$, $r \cdot t_1$ has principal part $r \cdot \underline{b}_1$.

Proof. Construct $\ell : D(2) \to V$ by

$$\ell(d_1, d_2) = \underline{a} + d_1 \cdot \underline{b}_1 + d_2 \cdot \underline{b}_2 .$$

Then $\ell \circ \mathrm{incl}_i = t_i \quad (i = 1,2)$, so that

$$\forall d \in D: \ (t_1 + t_2)(d) = \ell(d,d) = \underline{a} + d \cdot b_1 + d \cdot \underline{b}_2$$

$$= \underline{a} + d \cdot (\underline{b}_1 + \underline{b}_2)$$

The first result follows. The second is trivial.

One may express Axiom 1 for V by saying that, for each $\underline{a} \in V$, there is a canonical identification of $T_{\underline{a}} V$ with V, via principal-part formation. Proposition 5.2 expresses that this identification preserves the R-module structure, or equivalently that the isomorphism from Axiom 1 (for V)

$$V \times V \xrightarrow{\ \alpha\ } V^D$$

is an isomorphism of vector bundles over V (where the structural maps to the base space are, respectively, proj_1 and π). The composite γ

$$\gamma = V^D \xrightarrow{\ \cong\ } V \times V \xrightarrow{\ \mathrm{proj}_2\ } V \qquad (7.4)$$

associates to a tangent vector its principal part, and restricts to an R-linear map in each fibre, by the Proposition.

EXERCISES

7.1. The tangent bundle construction may be iterated. Construct a non trivial bijective map from $T(TM)$ to itself. Hint: $T(TM) = (M^D)^D \cong M^{D\times D}$; now use the "twist" map $D \times D \to D \times D$.

7.2. Assume that M is infinitesimally linear, so that

$$TM \times_M TM \cong M^{D(2)} \tag{7.5}$$

(cf. Exercise 6.5). Use the inclusion $D(2) \subseteq D \times D$ to construct a natural map

$$T(TM) \to TM \times_M TM. \tag{7.6}$$

Because of (7.5) and $T(TM) \cong M^{D\times D}$, a right inverse ∇ of (7.6) may be viewed as a right inverse of the restriction map $M^{D\times D} \to M^{D(2)}$, i.e. as the process of completing a figure

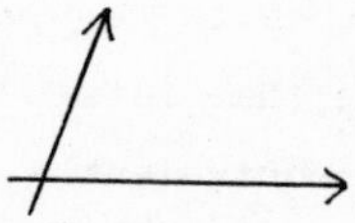

(a pair of tangent vectors at some point) into a figure

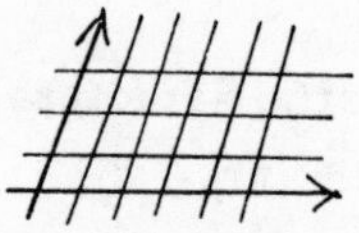

(a map $D \times D \to M$). Such ∇ thus is an infinitesimal notion of parallel transport: cf. [43].

7.3. Prove $(TM)^X \cong T(M^X)$. Note that this is almost trivial knowing that $TM = M^D$; for in any cartesian closed category,

$$(M^D)^X \cong (M^X)^D .$$

I.8: VECTOR FIELDS AND INFINITESIMAL TRANSFORMATIONS

The theory developed in the present § hopefully makes it clear why the cartesian closed structure of "the category $\mathcal{E}$ of smooth sets" is necessary to, and grows out of, natural physical/geometric considerations.

We noted in §7 that the tangent bundle of an object M was <u>representable</u> as the set M^D of maps from D to M. We quote from Lawvere [51], with slight change of notation:

"This representability of tangent (and jet) bundle functors by objects like D leads to considerable simplification of several concepts, constructions and calculations. For example, a first order ordinary differential equation, or vector field, on M is usually defined as a <u>section</u> $\hat{\xi}$ of the projection $\pi: M^D \to M, \ldots$" i.e.

$$M \xrightarrow{\hat{\xi}} M^D \quad \text{satisfying} \quad \pi \circ \hat{\xi} = id_M, \qquad (8.1)$$
$$\text{i.e. with} \quad \hat{\xi}(m)(0) = \forall m \in M$$

"But by the λ-conversion rule $\hat{\xi}$ is equivalent to a morphism

$$M \times D \longrightarrow M \quad \text{satisfying} \quad \xi(m,0) = m \quad \forall m \in M, \qquad (8.2)$$

i.e. to an "infinitesimal flow" of the additive group R". Also, by one further λ-conversion, we get

$$D \xrightarrow{\check{\xi}} M^M \quad \text{satisfying} \quad \check{\xi}(0) = id_M, \qquad (8.3)$$

i.e. an infinitesimal path in the space M^M of all transformations of M, or an infinitesimal deformation of the identity map. For fixed $d \in D$, the transformation $\check{\xi}(d) \in M^M$,

$$M \xrightarrow{\check{\xi}(d)} M,$$

is called an infinitesimal transformation of the vector field.

We shall see below that often (for instance when M is infinitesimally linear) the infinitesimal transformations of a vector field are bijective mappings $M \to M$, i.e. permute the elements of M, (cf. Corollary 8.2 below).

The presence of infinitesimal transformations $\check{\xi}(d)$ as actual transformations (permutations of the elements of M) is a feature the classical analytical approach to vector fields lacks, and which is indispensable for the natural synthetic reasonings with vector fields. When the analytic approach talks about "infinitesimal transformations" as synonymous with "vector fields", this is really unjustified, called for by a synthetic-geometric understanding which the formalism does not reflect: for a vector field does not, classically, permute anything; only the flows obtained by integrating the vector field (= ordinary differential equation) do that; and, even so, sometimes one cannot find any small interval $]-\varepsilon,\varepsilon[$ such that the flow can be defined on all of M with $]-\varepsilon,\varepsilon[$ as parameter interval, cf. Exercise 8.7.

The use of synthetic considerations about vector fields, in terms of their infinitesimal transformations as actual permutations, was used extensively by Sophus Lie. I am convinced also that the Lie bracket of vector fields (cf. §9 below) was conceived originally in terms of group theoretic commutators of infinitesimal transformations, but this I have not been able to document.

In the following, we will call any of the three equivalent data (8.1), (8.2), and (8.3) a vector field on M; also we will not always be pedantic whether to write $\hat{\xi}$, ξ, or $\check{\xi}$, in fact, we will prefer to use capital Latin letters like $X, Y, \ldots$, for vector fields.

Proposition 8.1. Assume M is infinitesimally linear. For any vector field $X: M \times D \to M$ on M, we have (for any $m \in M$)

$$\forall (d_1,d_2)\in D(2) \; : \; X(X(m,d_1),d_2) = X(m,d_1+d_2) \tag{8.4}$$

<u>Proof</u>. Note that the right hand side makes sense, since $d_1+d_2 \in D$ for $(d_1,d_2)\in D(2)$. Both sides in the equation may be viewed as functions $\ell: D(2) \to M$, and they agree when composed with $incl_1$ or $incl_2$, e.g. for $incl_2$:

$$X(X(m,0),d_2) = X(m,d_2) = X(m,0+d_2).$$

This proves the Proposition. Note that we only used the uniqueness assertion in the infinitesimal-linearity assumption.

The Proposition justifies the name "infinitesimal flow of the additive group R", because a global flow on M would traditionally be a map $\bar{X}: M\times R \to M$ satisfying (for any $m\in M$)

$$\bar{X}(\bar{X}(m,r_1),r_2) = \bar{X}(m,r_1+r_2) \tag{8.5}$$

for any $r_1,r_2\in R$, (as well as $\bar{X}(m,0) = m$).

<u>Corollary 8.2</u>. Assume M is infinitesimally linear. For any vector field X on M, we have

$$\forall d\in D: X(X(m,d),-d) = m.$$

In particular, each infinitesimal transformation $\check{X}(d): M \to M$ is invertible, with $\check{X}(-d)$ as inverse.

For any M, the set of vector fields on M is in an evident way a module over the ring R^M of all functions $M \to R$. For, if X is a vector field and $f: M \to R$ is a function, we define $f\cdot X$ by

$$(f\cdot X)(m,d) := X(m,f(m)\cdot d),$$

in other words, by multiplying the field vector $X(m,-)$ at m

with the scalar $f(m) \in R$.

Similarly, if M is infinitesimally linear, we can add two vector fields X and Y on M by adding, for each $m \in M$, the field vectors $X(m,-)$ and $Y(m,-)$ at m. By applying Proposition 7.2 pointwise, we immediately see, then:

Proposition 8.3. If M is infinitesimally linear, the set of vector fields on it is in a natural way a module over the ring R^M of R-valued functions on M.

EXERCISES

8.1. Prove that a map $\bar{X}: M \times R \to M$ is a flow in the sense of satisfying (8.5) and $\bar{X}(m,0) = m$ if and only if its exponential adjoint

$$R \longrightarrow M^M$$

is a homomorphism of monoids (the monoid structure on R being addition, and that of M^M composition of maps $M \to M$).

8.2. (Lawvere). Given objects M and N equipped with vector fields $X: M \times D \to M$ and $Y: N \times D \to N$, respectively. A map $f: M \to N$ is called a homomorphism of vector fields if

$$\begin{array}{ccc} M \times D & \xrightarrow{f \times D} & N \times D \\ {\scriptstyle X}\downarrow & & \downarrow{\scriptstyle Y} \\ M & \xrightarrow[f]{} & N \end{array}$$

commutes. Objects-equipped-with-vectorfields are thus organized into a category. Let $\frac{\partial}{\partial x}$ denote the vector field on R given by $\frac{\partial}{\partial x}(x,d) = x+d$. Prove (assuming Axiom 1) that a map $f: R \to R$ is an endomorphism of this object iff $f' \equiv 1$.

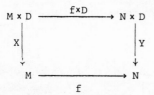

8.3. Assume M satisfies the Symmetric Functions Property (4.5), as well as Property W. Let X be a vector field on M. Prove that $\check{X}(d_1)$ commutes with $\check{X}(d_2)$ $\forall(d_1,d_2) \in D \times D$. Prove that we may extend $\check{X}: D \to M^M$ to $\check{X}_n: D_n \to M^M$ in such a way that the diagram

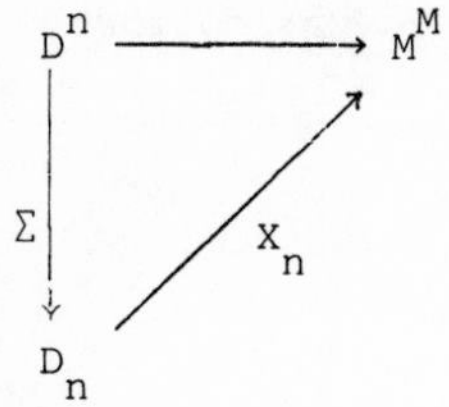

commutes, where the top map is

$$(d_1,\ldots,d_n) \longmapsto \check{X}(d_1) \circ \ldots \circ \check{X}(d_n).$$

Prove that the restriction of $\check{X}_{n+1}$ to D_n is $\check{X}_n$, and hence that we get a well defined map $\check{X}_\infty: D_\infty \to M^M$ having the various $\check{X}_n$'s as restrictions.

Prove that $\check{X}_\infty$ is a homomorphism of monoids (D_∞ with addition as monoid structure).

Thus $\check{X}_\infty$ is a flow in the sense that the equation (8.5) is satisfied for the exponential adjoint $X_\infty: M \times D_\infty \to M$ of $\check{X}_\infty$ (for $r_1, r_2 \in D_\infty$). The process $X \mapsto X_\infty$ described here is in essence equivalent to integration of the differential equation X by formal power series.

8.4. (Lawvere [50]). Let E_s be the subcategory of objects in E which satisfy Symmetric Functions Property and Property W; if $R \in E_s$, then so does D^n, D_n and D , by Exercise 6.7. Prove that $D^n/n! \cong D_n$, where $n!$ denotes the symmetric group in n letters, and $D^n/n!$ denotes its orbit space in E_s (a certain finite colimit in E_s). Reformulate the result of Exercise 8.3 by saying that

$$D_\infty = \sum_n D^n/n! \quad ("= e^D")$$

is the free monoid in E_s generated by the pointed set $(D,0)$ (the "sum" here is ascending union, rather than disjoint sum); and it is commutative.

8.5. Express the conclusion of Corollary 8.2 as follows: for any vector field X on M

$$(-X)^\vee(d) = X^\vee(-d) = (X^\vee(d))^{-1},$$

where we use the $^\vee$-notation as in (8.3).

8.6. Consider the map $\lambda: D \times D \to D$ given by $(d_1,d_2) \longmapsto (d_1+d_1\cdot d_2)$ (cf. Exercise 6.1). It induces a map

$$M^\lambda: M^D \to M^{D\times D} = (M^D)^D.$$

Prove that M^λ, via the displayed isomorphism, is a vector field on M^D. (This is the Liouville vector field considered in analytical mechanics, cf. [19] IX.2.4.)

8.7 (Classical calculus). Prove that the differential equation $y' = y^2$ has the property that there does not exist any interval $]-\varepsilon,\varepsilon[$ $(\varepsilon>0)$ such that, for each $x \in \mathbb{R}$, the unique solution $y(t)$ with $y(0) = x$ can be extended over the interval $]-\varepsilon,\varepsilon[$. (For, the solution is $y(t) = -1/(t-x^{-1})$, and for $x > 0$, say, this solution does not extend for $t \geq x^{-1}$.)

This can be reinterpreted as saying that the vector field $x^2 \frac{\partial}{\partial x}$ is not the "limit case" of any flow on $\mathbb{R}$; so there are <u>no</u> finite transformations $\mathbb{R} \to \mathbb{R}$ giving rise to the "infinitesimal transformation" $x^2 \frac{\partial}{\partial x}$.

(Compare e.g. [74] Vol.I Ch.5 for the classical connection between vector fields and differential equations.)

I.9: LIE BRACKET - COMMUTATOR OF INFINITESIMAL TRANSFORMATIONS

In this §, M is an arbitrary object which is infinitesimally linear and has the Property W:

> For any $\tau: D \times D \to M$ with $\tau(d,0) = \tau(0,d) = \tau(0,0)$ $\forall d \in D$, there exists a unique $t: D \to M$ with
> $\tau(d_1,d_2) = t(d_1 \cdot d_2)$ $\quad \forall (d_1,d_2) \in D \times D.$

(the same as in Exercise 4.2).

Assume that X and Y are vector fields on M; for each $(d_1,d_2) \in D \times D$, we consider the group theoretic commutator of the infinitesimal transformations $\check{X}(d_1)$ and $\check{Y}(d_2)$, i.e.

$$\check{Y}(-d_2) \circ \check{X}(-d_1) \circ \check{Y}(d_2) \circ \check{X}(d_1) \tag{9.1}$$

(utilizing Corollary 8.2: $\check{X}(-d)$ is the inverse of $\check{X}(d)$, and similarly for $\check{Y}$). If $d_1 = 0$, $\check{X}(d_1) = \check{X}(-d_1) = id_M$, so that (9.1) is itself id_M. Similarly if $d_2 = 0$. Thus (9.1) describes a map

$$D \times D \xrightarrow{\ \tau\ } M^M$$

with $\tau(0,d) = \tau(d,0) = id_M$ $\forall d \in D$. Since M^M has property W if M has (cf. Exercise 6.6), there exists a unique $t: D \to M^M$ with

$$t(d_1 \cdot d_2) = \tau(d_1,d_2) = \check{Y}(-d_2) \circ \check{X}(-d_1) \circ \check{Y}(d_2) \circ \check{X}(d_1),$$

$\forall (d_1,d_2) \in D \times D$. Clearly $t(0) = id_M$, so that t under the λ-conversion

$$\frac{D \longrightarrow M^M}{D\times M \longrightarrow M}$$

(cf. (8.3)-(8.2)) corresponds to a vector field $M\times D \longrightarrow M$ which we denote $[X,Y]$. Thus the characterizing property of $[X,Y]$ is that $\forall m \in M$, $\forall (d_1,d_2) \in D\times D$

$$[X,Y](m,d_1\cdot d_2) = Y(X(Y(X(m,d_1),d_2),-d_1),-d_2),$$

which in turn can be rigourously represented by means of a geometric figure (the names n,p,q and r for the four "new" points are for later reference)

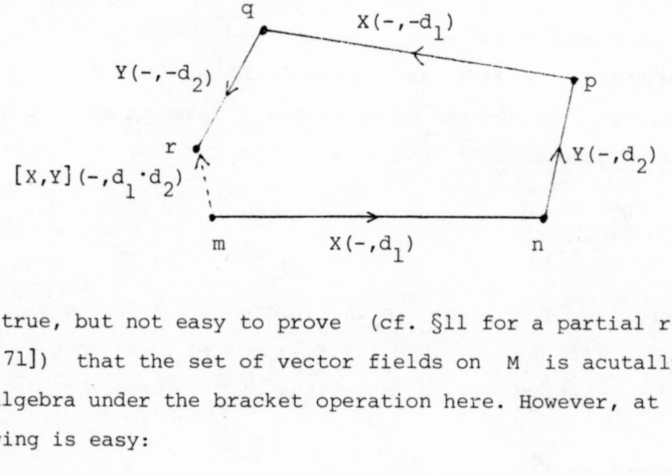

(9.3)

It is true, but not easy to prove (cf. §11 for a partial result, and [71]) that the set of vector fields on M is acutally an R-Lie algebra under the bracket operation here. However, at least the following is easy:

Proposition 9.1. For any vector fields X and Y on M,

$$[X,Y] = -[Y,X].$$

Proof. For any $(d_1,d_2) \in D\times D$

$$[X,Y]^\vee(d_1\cdot d_2) = \check{Y}(-d_2)\circ \check{X}(-d_1)\circ \check{Y}(d_2)\circ \check{X}(d_1)$$

$$= (\check{X}(-d_1)\circ \check{Y}(-d_2)\circ \check{X}(d_1)\circ \check{Y}(d_2))^{-1},$$

by $\check{X}(d_1)^{-1} = \check{X}(-d_1)$, and similarly for $\check{Y}$, and using the standard group theoretic identity $(b^{-1}a^{-1}ba)^{-1} = a^{-1}b^{-1}ab$,

$$= ([Y,X]^{\vee}(d_2 \cdot d_1))^{-1}$$

$$= (-[Y,X])^{\vee}(d_1 \cdot d_2),$$

by Corollary 8.2 (in the formulation of Exercise 8.5). Since this holds for all $(d_1,d_2) \in D \times D$, we conclude from the uniqueness assertion in Property W for M^M that $[X,Y]^{\vee} = (-[X,Y])^{\vee}$, whence the conclusion.

In classical treatments, to describe the geometric meaning of the Lie bracket of two vector fields, one first has to integrate the two vector fields into flows, then make a group-theoretic commutator of two transformations from the flows, and then pass to the limit, i.e. differentiate; cf. e.g. [61] §2.4 (in particular p.32). In the synthetic treatment, we don't have to make the detour of first integrating, and then differentiating.

Alternatively, the classical approach resorts to functional analysis identifying vector fields with differential operators, thereby abandoning the immediate geometric content like figure (9.3). The 'differential operators' associated to vector fields are considered in the next §.

EXERCISES

In the following exercises, M and G are objects that are infinitesimally linear and have Property W; R is assumed to satisfy Axiom 1.

9.1. Let X and Y be vector fields on M, and let $(d_1,d_2) \in D \times D$. Prove that the group theoretic commutator of $\check{X}(d_1)$ and $\check{Y}(d_2)$ equals that of $\check{X}(d_2)$, $\check{Y}(d_1)$. Also, prove that $\check{X}(d)$ commutes with $\check{Y}(d)$ for any $d \in D$.

9.2. Assume G has a group structure. A vector field X on G is called <u>left invariant</u> if for any $g_1, g_2 \in G$, and $d \in D$

$$g_1 \cdot X(g_2, d) = X(g_1 \cdot g_2, d).$$

Prove that the left-invariant vector fields on G form a sub-Lie-algebra of the Lie algebra of all vector fields.

9.3. Let G be as in Exercise 9.2, and let $e \in G$ be the neutral element. Prove that if $t \in T_eG$, then the law

$$X(g,d) := g \cdot t(d)$$

defines a left invariant vector field on G, and that this establishes a bijective correspondence between the set of left invariant vector fields on G, and T_eG. In particular, T_eG inherits a Lie algebra structure.

9.4. Generalize Exercises 9.2 and 9.3 from groups to monoids. Prove that the Lie algebra $T_e(M^M)$ may be identified with the Lie algebra of vector fields on M. In particular, general properties about Lie structure for vector fields may be reduced to properties of the Lie algebra T_eG. (This is the approach of [71].)

9.5. Let X and Y be vector fields on M. Prove that the following conditions are equivalent (and if they hold, we say that X and Y <u>commute</u>)

$$[X,Y] = 0 \qquad \text{(i)}$$

any infinitesimal transformation of the vector field X commutes with any infinitesimal transformation of the vector field Y (ii) any infinitesimal transformation of the vector field X is an endomorphism of the object (M,Y) in the catetory of vector fields (cf. Exercise 8.2 for this terminology) (iii)

9.6 (Lie [53]; cf. [34]). Let X and Y be vector fields on M and assume each field vector of X is an injective map $D \to M$ (X is a proper vector field). Also, we say m_1 and m_2 in M are X-neighbours if there exists $d \in D$ (necessarily unique) such that $X(m_1,d) = m_2$. Prove that the following conditions are equivalent (and if they hold, we say that X admits Y)

$$[X,Y] = \rho \cdot X \quad \text{for some} \quad \rho\colon \ M \to R \qquad \text{(i)}$$

the infinitesimal transformations of Y preserve the X-neighbour relation. (ii)

I.10: DIRECTIONAL DERIVATIVES

In this § we assume that R satisfies Axiom 1; V is assumed to be an R-module satisfying the vector form of Axiom 1, the most imporant case being of course $V = R$.

Let M be an object, and $X: M \times D \to M$ a vector field on it. For any function $f: M \to V$, we define $X(f)$, the <u>directional derivative</u> of f in the direction of the vector field, by (for fixed $m \in M$)

$$f(X(m,d)) = f(m) + d \cdot X(f)(m) \qquad \forall d \in D \tag{10.1}$$

This defines it uniquely, by applying Axiom 1 to the map $D \to V$ given by $f(X(m,-))$.

Diagrammatically, $X(f)$ is the composite

$$M \xrightarrow{\hat{X}} M^D \xrightarrow{f^D} V^D \xrightarrow{\gamma} V \tag{10.2}$$

where $\gamma(t)$ = principal part of t = the unique $\underline{b} \in V$ such that $t(d) = t(0) + d \cdot \underline{b} \quad \forall d \in D$.

Consider in particular $M = R$, $V = R$, and the vector field

$$\frac{\partial}{\partial x} : R \times D \longrightarrow R$$

given by $(x,d) \longmapsto x+d$. Then the process $f \mapsto X(f)$ is just the differentiation $f \mapsto f'$ described in §2. The rules proved there immediately generalize; thus we have

Theorem 10.1. For any $f, g\colon M \to V$, $r \in R$, and $\varphi\colon M \to R$, we have

$$X(r\cdot f) = r\cdot X(f) \qquad \text{(i)}$$

$$X(f+g) = X(f) + X(g) \qquad \text{(ii)}$$

$$X(\varphi\cdot f) = X(\varphi)\cdot f + \varphi\cdot X(f) \qquad \text{(iii)}$$

Clearly $X(f) \equiv 0$ if f is constant. More generally, a function $f\colon M \to V$ such that $X(f) \equiv 0$ is called an integral or a first integral of X. Clearly, by (10.1), $X(f) \equiv 0$ iff for all $m \in M$ and $d \in D$, $f(X(m,d)) = f(m)$. This condition can be reformulated:

$$f \circ \check{X}(d) = f,$$

in other words, f is invariant under the infinitesimal transformations of the vector field X. (This, in turn, might suggestively be expressed: "f is constant on the orbit of the action of X".)

The following result will be useful in stating and proving linearity conditions. It is not so surprising, since in classical calculus, the corresponding result holds for smooth functions between coordinate vector spaces.

Proposition 10.2. Let U and V be R-modules (V satisfying Axiom 1). Then any map $f\colon U \to V$ satisfying the "homogeneity" condition

$$\forall r \in R\ \forall \underline{u} \in U\colon f(r\cdot\underline{u}) = r\cdot f(\underline{u})$$

is R-linear.

Proof. For $\underline{y} \in U$, we denote by $D_{\underline{y}}$ the vector field $U \times D \to U$ given by

$$D_{\underline{y}}(\underline{u},d) = \underline{u} + d\cdot\underline{y} \quad ;$$

in particular, for $g\colon U \to V$, we have as a special case of (10.1)

$$g(\underline{u}+d\cdot\underline{y}) = g(\underline{u}) + d\cdot D_{\underline{y}}(g)(\underline{u}), \qquad \forall d \in D.$$

In particular, for $d \in D$,

$$d\cdot f(\underline{x}+\underline{y}) = f(d\cdot\underline{x} + d\cdot\underline{y})$$

$$= f(d\cdot\underline{x}) + d\cdot D_{\underline{y}} f(d\cdot\underline{x})$$

$$= f(d\cdot\underline{x}) + d\cdot(D_{\underline{y}} f(\underline{0}) + d\cdot D_{\underline{x}} D_{\underline{y}} f(\underline{0}))$$

$$= f(d\cdot\underline{x}) + d\cdot D_{\underline{y}} f(0)$$

(since $d^2 = 0$)

$$= f(d\cdot\underline{x}) + f(d\cdot\underline{y})$$

$$= d\cdot f(\underline{x}) + d\cdot f(\underline{y}),$$

the first and the last equality sign by homogeneity condition. Since this holds for all $d \in D$, we get the additivity of f by cancelling the universally quantified d. Thus f is R-linear.

Note that to prove additivity, only homogeneity conditions for scalars in D was assumed. This observation is utilized in Exercise 10.2.

In the following theorem, we assume that M is infinitesimally linear and has Property W.

Theorem 10.3. For any vector fields X, X_1, X_2, Y on M, any $\varphi\colon M \to R$, and any $f\colon M \to V$, we have

$$(\varphi\cdot X)(f) = \varphi\cdot X(f) \tag{i}$$

$$(X_1+X_2)(f) = X_1(f) + X_2(f) \tag{ii}$$

$$[X,Y]\,(f) = X(Y(f)) - Y(X(f)) \tag{iii}$$

Proof. Using the definition of $\varphi\cdot X$, and (10.1), we have, for $\forall m \in M$, $\forall d \in D$:

$$f((\varphi\cdot X)(m,d)) = f(X(m,\varphi(m)\cdot d)) = f(m) + (\varphi(m)\cdot d) \bullet X(f)(m)$$

(noting that $\varphi(m)\cdot d \in D$). On the other hand, directly by (10.1)

$$f((\varphi\cdot X)(m,d)) = f(m) + d\cdot(\varphi\cdot X)(f)(m).$$

Comparing these two equations, and cancelling the universally quantified d, we get $\varphi(m)\cdot X(f)(m) = (\varphi\cdot X)(f)(m)$, proving (i). To prove (ii), let $f: M \to V$ be fixed. The process

$$X \longmapsto X(f)$$

is a map $g: \mathrm{Vect}(M) \to V^M$, where $\mathrm{Vect}(M)$ is the set of vector fields on M; $\mathrm{Vect}(M)$ is an R-module, since M is infinitesimally linear. Also, V^M satisfies the vector form of Axiom 1 since V does. From (i) follows that $g(r\cdot X) = r\cdot g(X)\ \forall r \in R$, and (ii) then follows from Proposition 10.2.

Let us finally prove (iii). For fixed m, d_1, d_2, we consider the circuit (9.3) and the elements n,p,q,r described there. We consider $f(r) - f(m)$. First

$$f(r) = f(q) - d_2\cdot Y(f)(q)$$

$$= f(p) - d_1\cdot X(f)(p) - d_2\cdot Y(f)(q)$$

using the "generalized Taylor formula" (10.1) twice. Again, using generalized Taylor twice, (noting $m = X(n,-d_1)$, and $n = Y(p,-d_2)$, by Corollary 8.2)

$$f(m) = f(n) - d_1 \cdot X(f)(n)$$

$$= f(p) - d_2 \cdot Y(f)(p) - d_1 \cdot X(f)(n).$$

Subtracting these two equations, we get

$$f(r) - f(m) = d_1 \cdot \{X(f)(n) - X(f)(p)\} + d_2 \cdot \{Y(f)(p) - Y(f)(q)\}$$

$$= -d_1 \cdot d_2 \cdot Y(X(f))(p) + d_1 \cdot d_2 \cdot X(Y(f))(p), \qquad (10.3)$$

using generalized Taylor on each of the curly brackets. Now we have, for any $g: M \to V$,

$$d_2 \cdot g(p) = d_2 \cdot g(n),$$

since

$$d_2 \cdot g(p) = d_2 \cdot g(Y(n,d_2)) = d_2 \cdot (g(n) + d_2 \cdot Y(g)(n)),$$

and using $d_2^2 = 0$. Similarly we have

$$d_1 \cdot g(n) = d_1 \cdot g(m),$$

so that, combining these two equations, we have

$$d_1 \cdot d_2 \cdot g(p) = d_1 \cdot d_2 \cdot g(m).$$

Applying this for $g = Y(X(f))$ and $g = X(Y(f))$, we see that the argument p in (10.3) may be replaced by m; so (10.3) is replaced by

$$f(r) - f(m) = d_1 \cdot d_2 \cdot (X(Y(f))(m) - Y(X(f))(m)). \qquad (10.4)$$

On the other hand

$$[X,Y](m,d_1 \cdot d_2) = r,$$

so that, by generalized Taylor,

$$f(r) - f(m) = d_1 \cdot d_2 \cdot [X,Y](f)(m). \tag{10.5}$$

Comparing (10.4) and (10.5), we see that

$$d_1 \cdot d_2 \cdot (X(Y(f))(m) - Y(X(f))(m)) = d_1 \cdot d_2 \cdot [X,Y](f)(m),$$

and since this holds for all $(d_1,d_2) \in D \times D$, we may cancel the d_1 and d_2 one at a time, to get (iii).

EXERCISES

10.1. Let $\frac{\partial}{\partial x_i}$ (for $i = 1,\ldots,n$) denote the vector field on R^n given, as a map $R^n \times D \to R^n$, by

$$((x_1,\ldots,x_n),d) \longmapsto (x_1,\ldots,x_i+d,\ldots,x_n).$$

(i) Prove that $\frac{\partial}{\partial x_i}$ commutes with $\frac{\partial}{\partial x_j}$ (terminology of Exercise 9.6).

(ii) Prove that directional derivative along $\frac{\partial}{\partial x_i}$ equals the i'th partial derivative (§4). Note that (i) makes sense, and is easy to prove, without any consideration of directional derivation, in fact does not even depend on Axiom 1.

10.2 (Veit.) Assume that V has the property that any $g: R \to V$ with $\frac{\partial g}{\partial x}$ $(= g') \equiv 0$ is constant. Prove that the assumption in Proposition 10.2 can be weakened into

$$\forall d \in D \ \ \forall \underline{u} \in U: f(d \cdot \underline{u}) = d \cdot f(\underline{u}).$$

(Hint: to prove the homogeneity condition for arbitrary scalars $r \in R$, consider, for fixed $\underline{u}$, the function $g(r) := f(r\underline{u}) - r \cdot f(\underline{u})$.)

10.3. For X a vector field on M, and $f: M \to V$ a function, express the condition $X(f) \equiv 0$ as the statement: f is a morphism in the category of objects-with-a-vectorfield from (M,X) to $(V,0)$.

I.11: SOME ABSTRACT ALGEBRA AND FUNCTIONAL ANALYSIS. APPLICATION TO PROOF OF JACOBI IDENTITY

Recall that an R-algebra C is a commutative ring equipped with a ring map $R \to C$ (which implies an R-module structure on C). The ring R^M of all functions from M to R (M an arbitrary object) is an evident example.

Recall also that if C_1 and C_2 are R-algebras, and $i: C_1 \to C_2$ is an R-algebra map, an R-derivation from C_1 to C_2 (relative to i) is an R-linear map

$$\delta: C_1 \to C_2$$

such that

$$\delta(c_1 \cdot c_2) = \delta(c_1) \cdot i(c_2) + i(c_1) \cdot \delta(c_2) \quad \forall c_1, c_2 \in C.$$

They form in an evident way an R-module, denoted $\mathrm{Der}_R^i(C_1, C_2)$. If $C_1 = C_2 = C$ and i = identity map, we just write $\mathrm{Der}_R(C,C)$.

It is well known, and easy to see, that, for C an R-algebra, the R-module

$$\mathcal{D} = \mathrm{Der}_R(C,C)$$

has a natural structure of <u>Lie algebra over</u> R, meaning that there is an R-bilinear map

$$\mathcal{D} \times \mathcal{D} \xrightarrow{[-,-]} \mathcal{D} \qquad (11.1)$$

(given here by

$$[\delta_1,\delta_2] = \delta_1 \circ \delta_2 - \delta_2 \circ \delta_1\ )$$

such that $[-,-]$ satisfies the Jacobi identity

$$[\delta_1,[\delta_2,\delta_3]] + [\delta_2,[\delta_3,\delta_1]] + [\delta_3,[\delta_1,\delta_2]] = 0 \qquad (11.2)$$

as well as

$$[\delta_1,\delta_2] + [\delta_2,\delta_1] = 0 \qquad (11.3)$$

for all $\delta_1,\delta_2,\delta_3 \in \mathcal{D}$ (trivial verification, but for (11.2), not short).

Also there is a multiplication map

$$C \times \mathcal{D} \xrightarrow{\quad \cdot \quad} \mathcal{D} \qquad (11.4)$$

as well as an evaluation map

$$\mathcal{D} \times C \xrightarrow{\quad -(-) \quad} C, \qquad (11.5)$$

due to the fact that $\mathcal{D}$ is a set of functions $C \to C$. Both these maps are R-bilinear, and furthermore, for all $\delta_1,\delta_2 \in \mathcal{D}$ and $c \in C$, we have

$$[\delta_1, c\cdot\delta_2] = \delta_1(c)\cdot\delta_2 + c\cdot[\delta_1,\delta_2] \qquad (11.6)$$

The R-bilinear structures (11.1), (11.4), and (11.5) form what is called an R-Lie-module, (more precisely, they make $\mathcal{D}$ into a Lie module over C), cf. [61] §2.2. (The defining equations for this notion are (11.2), (11.3), and (11.6).)

We now consider in particular $C = R^M$, where R is assumed to satisfy Axiom 1, and M is assumed to be infinitesimally linear and have Property W. By Theorem 10.1 we then have a map

$$\mathrm{Vect}(M) \longrightarrow \mathcal{D} = \mathrm{Der}_R(R^M,R^M), \qquad (11.7)$$

where $\mathrm{Vect}(M)$ is the R^M-module of vector fields on M, given by

$$X \longmapsto [f \longmapsto X(f)].$$

This map is R^M-linear, by Theorem 10.3 (i) and (ii), and (iii) tells us that it preserves the bracket operation.

Theorem 11.1. If the map (11.7) is injective, then $\mathrm{Vect}(M)$, with its R^M-module structure, and the Lie bracket of §9, becomes an R-Lie algebra. It becomes in fact a Lie module over R^M, by letting the evaluation map $\mathrm{Vect}(M) \times R^M \longrightarrow R^M$ be the formation of directional derivative, $(X,f) \longmapsto X(f)$.

Proof. We just have to verify the equations (11.2), (11.3), and (11.6); (11.2) and (11.3) follow, because (11.7) preserves the R-linear structure and the bracket; to prove (11.6) means to prove for $X_1, X_2 \in \mathrm{Vect}(M)$ and $\varphi: M \longrightarrow R$

$$[X_1, \varphi\cdot X_2] = X_1(\varphi)\cdot X_2 + \varphi\cdot[X_1,X_2]. \tag{11.8}$$

By injectivity of (11.7), it suffices to prove for arbitrary $f: M \to R$

$$[X_1,\varphi\cdot X_2](f) = (X_1(\varphi)\cdot X_2)(f) + (\varphi\cdot[X_1,X_2])(f),$$

but this is an immediate calculation based on Theorem 10.3, and Theorem 10.1 (iii).

Note that (11.3) was proved also in §9 (Proposition 9.1) by "pure geometric methods"; also, the Jacobi identity (11.2) can be proved this way (cf. [71]), using P.Hall's 42-letter identity in the theory of groups. I don't know whether (11.8) can be proved without resort to functional analysis, i.e. without resort to (11.7), except in case where M is an infinitesimally linear R-module satisfying Axiom 1, or if M is a "formal manifold" in the sense of §17 below.*

Classically, for M a finite dimensional manifold, the analogue of the map (11.7) is bijective, so that vector fields here are identified by the differential operators to which they give rise. This has computational advantages, but the geometric content ("infinitesimal transformations") seems lost.

We interrupt here the naive exposition in order to present a strong comprehensive axiom (of functional-analytic character), which at the same time will throw some light (cf. Proposition 12.2) on the question when the ring R^M is big enough to recover M, Vect(M), etc. Also the assumption of (11.7) being injective is an assumption in the same spirit. For a more precise statement cf. Corollary 12.5.

EXERCISES

11.1. In any group G, let $\{x,y\} := x^{-1}y^{-1}xy$, and $x^y := y^{-1}xy$. Prove

$$\{x^y,\{y,z\}\}\cdot\{y^z,\{z,x\}\}\cdot\{z^x,\{x,y\}\} = e \quad \text{(P.Hall's identity)}.$$

---

* (Added in proof). R. Lavendhomme has now given a geometric proof of (11.8), just assuming infinitesimal linearity and Condition W for M.

I.12: THE COMPREHENSIVE AXIOM

Most of the present § is not required in the rest of Part I, except in Theorem 16.1. The reader who wants to skip over to §13, may in §16 take the conclusion ("Axiom 1^W") of Theorem 16.1 as his "comprehensive axiom" instead.

Let k be a commutative ring in Sets.

A finitely presented k-algebra is a k-algebra B of form

$$B = k[X_1,\ldots,X_n]/(f_1(X_1,\ldots,X_n),\ldots,f_m(X_1,\ldots,X_n)) \qquad (12.1)$$

where the f_i's are polynomials with k-coefficients. Examples:

$$k[X] \qquad (i)$$

$$k[X]/(X^2) \quad \text{(denoted } k[\varepsilon]) \qquad (ii)$$

$$k[X,Y]/(X^2+Y^2-1). \qquad (iii)$$

(Note: B is a ring in the category Set). Each such presentation (12.1) should be viewed as a prescription for carving out a sub-"set" of C^n for any commutative k-algebra (-object) C in any category E with finite inverse limits. The "subsets carved out" by (i), (ii), and (iii) are

$$C \text{ itself} \qquad (i')$$

$$[\![c \in C \mid c^2 = 0]\!] \text{ (which, if } C = R, \text{ is just } D) \qquad (ii')$$

$$[\![(c_1,c_2) \in C \times C \mid c_1^2 + c_2^2 - 1 = 0]\!], \text{ the "unit circle" in } C \times C. \qquad (iii')$$

We denote the object carved out from C^n by B by the symbol $\mathrm{Spec}_C(B)$. The notation is meant to suggest that we (for fixed C) associate a "geometric" object in $\mathcal{E}$ to the algebraic object B (e.g. to the ring B in (iii) associate "the circle" in (iii')). If we denote the category of finitely presented k-algebras by $\mathrm{FP}\mathbb{T}_k$, then Spec_C is a finite-inverse-limit-preserving functor $(\mathrm{FP}\mathbb{T}_k)^{op} \longrightarrow \mathcal{E}$ with $k[X] \longmapsto C$, and, taking the k-algebra structure on C suitably into account, this in fact determines it up to unique isomorphism, see Appendix A. Also, if $\mathcal{E}$ is cartesian closed,

$$\mathrm{Spec}_{R^X}(B) = (\mathrm{Spec}_R(B))^X,$$

for any $X \in \mathcal{E}$.

We henceforth in this § assume that $\mathcal{E}$ is cartesian closed and has finite inverse limits.

Then we can state the comprehensive Axiom for our commutative k-algebra object R in $\mathcal{E}$. (A related Axiom B.2 will be considered in III §9.)

<u>Axiom</u> 2_k. For any finitely presented k-algebra B, and any R-algebra C in $\mathcal{E}$, the canonical map (cf. below)

$$\underset{\sim}{\mathrm{hom}}_{R\text{-Alg}}(R^{\mathrm{Spec}_R(B)}, C) \xrightarrow{\nu_{B,C}} \mathrm{Spec}_C(B) \qquad (12.2)$$

is an isomorphism.

Here, $R^{\mathrm{Spec}_R(B)}$, like any R^M, is an R-algebra, and $\underset{\sim}{\mathrm{hom}}_{R\text{-Alg}}$ denotes the object (in $\mathcal{E}$) of R-algebra maps.

For the case where $B = k[X]$ so that $\mathrm{Spec}_R(B) = R$, $\mathrm{Spec}_C(B) = C$, the map $\nu_{B,C}$

$$\underset{\sim}{\mathrm{hom}}_{R\text{-Alg}}(R^R, C) \longrightarrow C \qquad (12.3)$$

is: evaluation at $\mathrm{id}_R \in R^R$. This determines it by Appendix A, (Theorem A.1). The map $\nu_{B,C}$ occurs, in a more general context, in (A.1) in Appendix A.

It is easy to see that if $k \to K$ is a ring-homomorphism, then for any K-algebra object R, Axiom 2_K implies Axiom 2_k.

We have

Theorem 12.1. Axiom 2_k implies Axiom 1 (as well as Axiom 1', Axiom 1",...).

We postpone the proof until §16, where we prove a stronger result (Theorem 16.1), involving the notion of Weil-algebra, to be described there.

We derive now some further consequences of Axiom 2_k. In the rest of this §, R is a k-algebra object in a cartesian closed category $\mathcal{E}$ with finite inverse limits, and is assumed to satisfy Axiom 2_k. We use set theoretic notation.

By an affine scheme (relative to R) we mean an object of form $\mathrm{Spec}_R(B)$ for some $B \in \mathrm{FP}\mathbb{T}_k$. We then have

Theorem 12.2. Affine schemes have the property that for any object $X \in \mathcal{E}$, the canonical map

$$M^X \xrightarrow{\eta} \underset{\sim}{\mathrm{hom}}_{R\text{-Alg}}(R^M, R^X)$$

given by

$$f \longmapsto [g \longmapsto g \circ f]$$

for $f: X \to M$ and $g: M \to R$, is an isomorphism.

Proof. For each $M = \mathrm{Spec}_R(B)$, we have the canonical map ν from (12.2), with $C = R^X$,

$$\underset{\sim}{\mathrm{hom}}_{R\text{-Alg}}(R^{\mathrm{Spec}_R(B)}, R^X) \xrightarrow{\nu} \mathrm{Spec}_{R^X}(B),$$

and it is an isomorphism, by Axiom 2. But $\mathrm{Spec}_{R^X}(B) \cong (\mathrm{Spec}_R(B))^X$. To see $\nu \circ \eta = \mathrm{id}_{M^X}$, it suffices, by naturality of both, and by Appendix A (Theorem A.1, last statement), to see this for the case $B = k[Y]$ (so $M = \mathrm{Spec}_R(k[Y]) = R$). For $f \in R^X = (\mathrm{Spec}_R(k[Y]))^X$,

$$\eta(f) = [g \longmapsto g \circ f]$$

and applying ν means 'evaluation at id_R'. So

$$\nu(\eta(f)) = \mathrm{id}_R \circ f = f.$$

Corollary 12.3. Each affine scheme M is reflexive in the sense that the canonical map to the double dual, relative to R,

$$M \longrightarrow \underset{\sim\sim}{\mathrm{hom}}_{R\text{-Alg}}(R^M, R)$$

(sending m to 'evaluation at m') is an isomorphism.

Proof. Take $X = \mathbf{1}$ in the Theorem.

Corollary 12.4. Each affine scheme M has the property that its tangent space $T_m N$ can be identified with the object of R-derivations from R^M to R relative to the R-algebra map ev_m ('evaluation at m').

Proof. In Theorem 12.2, take $X = D = \mathrm{Spec}_R(k[\varepsilon])$, and utilize $R^D \cong R[\varepsilon]$ (Corollary 1.2 of Axiom 1, which we may apply now, since Axiom 1 holds in virtue of Theorem 12.1). The theorem then gives an isomorphism

$$M^D \longrightarrow \underset{\sim\sim}{\mathrm{hom}}_{R\text{-Alg}}(R^M, R[\varepsilon]). \qquad (12.5)$$

Now an R-algebra map from an R-algebra C into $R[\varepsilon]$ is well known (cf. [23] §20) to be the same as a pair of R-linear maps $C \to R$, where the first is an R-algebra map and the second is a derivation relative to the first.

A related consequence is

Corollary 12.5. Each affine scheme M has the property that the comparison map (11.7)

$$\text{Vect}(M) \longrightarrow \text{Der}_R(R^M, R^M)$$

is an isomorphism.

Proof. The right hand side sits in a pull-back square

$$\begin{array}{ccc} \text{Der}_R(R^M,R^M) & \rightarrowtail & \underset{\sim}{\text{hom}}_{R\text{-Alg}}(R^M, R^M[\varepsilon]) \\ \big\downarrow & & \big\downarrow \\ \mathbf{1} & \xrightarrow[\ulcorner \text{id}_{R^M} \urcorner]{} & \underset{\sim}{\text{hom}}_{R\text{-Alg}}(R^M, R^M), \end{array}$$

where the right hand vertical map is induced by that R-algebra map $R^M[\varepsilon] \longrightarrow R^M$ which sends ε to 0. On the other hand, we have identifications (the middle one by Axiom 1):

$$R^M[\varepsilon] \cong (R[\varepsilon])^M \cong (R^D)^M \cong R^{M\times D} ,$$

and under these identifications, the right hand vertical map gets identified with the right hand vertical one in the commutative diagram

$$\begin{array}{ccccc} \text{Vect}(M) & \rightarrowtail & M^{M\times D} & \xrightarrow{\cong} & \underset{\sim}{\text{hom}}_{R\text{-Alg}}(R^M, R^{M\times D}) \\ \big\downarrow & & \big\downarrow & & \big\downarrow \\ \mathbf{1} & \xrightarrow[\ulcorner \text{id}_M \urcorner]{} & M^M & \xrightarrow[\cong]{} & \underset{\sim}{\text{hom}}_{R\text{-Alg}}(R^M, R^M) \end{array}$$

Here the vertical maps are induced by the map $M \to M \times D$ given by $m \mapsto (m,0)$, and the horizontal isomorphisms are those of Theorem 12.2. The left hand square is a pull-back. Thus $\text{Vect}(M) \cong \text{Der}_R(R^M, R^M)$. We leave to the reader to keep track of the identifications.

<u>Proposition 12.6</u>. The R-algebra R^R is the free R-algebra on one generator, namely $id_R \in R^R$.

(By this we mean that for any R-algebra C, the map "evaluation at id_M"

$$\underset{\sim}{\text{hom}}_{R\text{-Alg}}(R^R, C) \to C,$$

is an isomorphism.)

<u>Proof</u>. We have

$$\underset{\sim}{\text{hom}}_{R\text{-Alg}}(R^R, C) = \underset{\sim}{\text{hom}}_{R\text{-Alg}}(R^{\text{Spec}_R(k[X])}, C)$$

$$\cong \text{Spec}_C(k[X]) = C,$$

the isomorphism being a case of the ν of Axiom 2_k.

Note that if we let X denote the identity map of R, which is a standard mathematical practice, then the Proposition may be expressed

$$R^R = R[X].$$

<u>Proposition 12.7</u>. The functor $FP\mathbb{T}_k \to R\text{-Alg}$ given by

$$B \longmapsto R^{\text{Spec}_R(B)}$$

preserves finite colimits.

Proof. For any R-algebra C, and any finite colimit $\varinjlim_i (B_i)$ in $FP\mathbb{T}_k$, we have (writing Spec for Spec_R), by Axiom 2_k

$$\underset{\sim}{\mathrm{hom}}_{R\text{-Alg}}(R^{\mathrm{Spec}(\varinjlim B_i)}, C) \cong \mathrm{Spec}_C(\varinjlim B_i)$$

$$\cong \varprojlim \mathrm{Spec}_C(B_i)$$

(because $\mathrm{Spec}_C : (FP\mathbb{T}_k)^{op} \longrightarrow E$ is left exact); and then, by Axiom 2 again,

$$\cong \varprojlim \underset{\sim}{\mathrm{hom}}_{R\text{-Alg}}(R^{\mathrm{Spec}(B_i)}, C)$$

$$= \underset{\sim}{\mathrm{hom}}_{R\text{-Alg}}(\varinjlim R^{\mathrm{Spec}(B_i)}, C),$$

naturally in C. By Yoneda's lemma, and keeping track of the identifications, the result follows.

EXERCISES

12.1. It will be proved in §16 that Axiom 2_k implies that R has the properties: it is infinitesimally linear and has Property W as well as the Symmetric Function Property, (provided k is a $\mathbb{Q}$-algebra). Assuming this result, prove that Axiom 2_k implies that any reflexive object, and in particular, any affine scheme, has these three properties.

Hint: R^X has for any X, these properties. Now use that $\underset{\sim}{\mathrm{hom}}_{R\text{-Alg}}(R^M, R)$ is carved out of R^{R^M} by equalizers with values in R and R^R, and use Exercise 6.6.

12.2. Construct, on basis of Axiom 1 alone, a map like (12.5) (not necessarily an isomorphism). Use this to construct a map

$$M^D \longrightarrow \underset{\sim}{\mathrm{hom}}_{R\text{-lin}}(R^M, R) \times \underset{\sim}{\mathrm{hom}}_{R\text{-lin}}(R^M, R)$$

where $\underset{\sim}{\mathrm{hom}}_{R\text{-lin}}$ denotes the object of R-linear maps. Prove that if M is reflexive, then this map is injective.

12.3. Generalize Exercise 12.2: assuming Axiom 1', construct a map

$$M^{D_k} \longrightarrow \Pi \underset{\sim\sim}{\mathrm{hom}}_{R\text{-lin}}(R^M, R)$$

(k+1-fold product), and prove that if M is reflexive, this map is injective.

The object $\underset{\sim\sim}{\mathrm{hom}}_{R\text{-lin}}(R^M, R)$ should be considered the object of <u>distributions</u> on M with compact support. It reappears in §14.

I.13: ORDER AND INTEGRATION

The geometric line has properties and structure not taken into account in the preceding §'s, namely its ordering, and the possibility of integrating functions. The axiomatizations of these two things are best introduced together, even though it is possible to do it separately; thus, an obvious Axiom for integration would be to require existence of primitives (anti-derivatives):

$$\forall f: R \to R \quad \exists!\, g: R \to R \text{ with } g' \equiv f \text{ and } g(0) = 0 \qquad (*)$$

The the number $\int_a^b f(x)dx$ (say) would be defined as $g(b)-g(a)$. But to define this number, it should suffice for f to be defined on the interval $[a,b]$ only, not on the whole line. So (*) is too weak an axiom because it has too strong assumption on f. So, essentially, we want to have an axiom giving antiderivatives for functions $f: [a,b] \to R$ for $[a,b]$ any interval, and to define the notion of <u>interval</u>, we need to make explicit the ordering $\leq$ of the geometric line R. Besides the commutative ring structure on R, we consider therefore its 'order' relation $\leq$ which is assumed

transitive: $x \leq y \wedge y \leq z \Rightarrow x \leq z$,

reflexive: $x \leq x$,

and

compatible with the ring structure:

$$x \leq y \Rightarrow x+z \leq y+z$$

$$x \leq y \wedge 0 \leq t \Rightarrow x \cdot t \leq y \cdot t \qquad (13.1)$$

$$0 \leq 1.$$

Furthermore, we assume

$$d \text{ nilpotent} \Rightarrow 0 \le d \wedge d \le 0. \tag{13.2}$$

Note that we do not assume $\le$ to be antisymmetric ("$x \le y \wedge y \le x \Rightarrow x=y$"), because that would force all nilpotent elements to be 0, by (13.2). In other words, $\le$ is a preorder, not a partial order.

Intervals are then defined in the expected way:

$$[a,b] := [\![x \in R \mid a \le x \le b]\!] .$$

Note that a and b cannot be reconstructed from $[a,b]$, since for any nilpotent d, $[a,b] = [a,b+d]$, by (13.2). (For this reason, $[a,b]$ will in §15 be denoted $|[a,b]|$, to reserve the notation $[a,b]$ for something where the information of the end points is retained.)

Note also that, by (13.2), any interval $[a,b] = U$ has the property (2.3)

$$x \in [a,b] \wedge d \in D \Rightarrow x+d \in [a,b],$$

so that, if $g: [a,b] \to R$, then g' can be defined on the whole of $[a,b]$ (assuming Axiom 1, of course).

Finally, note that (13.1) implies that any interval is convex: $x,y \in [a,b] \wedge 0 \le t \le 1 \Rightarrow x+t\cdot(y-x) \in [a,b]$.

In the rest of this §, we assume such an ordering $\le$ on R; and we assume Axiom 1 as well as the

Integration Axiom. For any $f: [0,1] \to R$, there exists a unique $g: [0,1] \to R$ such that $g' \equiv f$ and $g(0) = 0$.

We can then define

$$\int_0^1 f(t)dt := g(1) \qquad (= g(1)-g(0)).$$

Several of the standard rules for integration then follow from the corresponding rules for differentiation (Theorem 2.2) purely formally. In particular, the process

$$f \longmapsto \int_0^1 f(t)dt$$

depends in an R-linear way on f, so defines an R-linear map

$$R^{[0,1]} \to R.$$

Also, for any $h\colon [0,1] \to R$,

$$\int_0^1 h'(t)dt = h(1) - h(0). \tag{13.3}$$

From these two properties, we can deduce

<u>Proposition 13.1</u> ("Hadamard's lemma")*. For any $a,b \in R$, any $f\colon [a,b] \longrightarrow R$, and any $x,y \in [a,b]$, we have

$$f(y) - f(x) = (y-x) \cdot \int_0^1 f'(x+t\cdot(y-x))dt.$$

<u>Proof</u>. (Note that the integrand makes sense because of convexity of $[a,b]$). For any $x,y \in [a,b]$, we have a map $\varphi\colon [0,1] \longrightarrow [a,b]$ given by $\varphi(t) = x + t \cdot (y-x)$. We have $\varphi' \equiv y-x$. So

$$\begin{aligned} f(y) - f(x) &= f(\varphi(1)) - f(\varphi(0)) \\ &= \int_0^1 (f \circ \varphi)'(t)dt && \text{(by (13.3))} \\ &= \int_0^1 (y-x)\cdot(f' \circ \varphi)(t)dt && \text{(chain rule, Theorem 2.2)} \\ &= (y-x) \cdot \int_0^1 (f' \circ \varphi)(t)dt, \end{aligned}$$

*For the categorical interpretation of this, and the rest of the §, we need that the base category is <u>stably</u> cartesian closed, cf. II §6.

the last equaltiy by linearity of the integration process. But this is the desired equality.

It is possible to prove several of the standard rules for integrals and antiderivatives, like "differentiating under the integral sign",... . Also, one may prove, essentially using the same technique as in the proof of Proposition 13.1, the following

Theorem 13.2. For any $a \leq b$ and any $f: [a,b] \longrightarrow R$, there is a unique $g: [a,b] \to R$ with $g' \equiv f$ and $g(a) = 0$.

We refer the reader to [44] for the proof. If f and g are as in the theorem, we may define $\int_a^b f(t)dt$ as $g(b)$. This is consistent with our previous definition of $\int_0^1 f(t)dt$.

We quote from [44] some results concerning $\int_a^b f(t)dt$:

$$\int_a^b f(t)dt \text{ depends in an R-linear way on } f \quad (\text{where } a \leq b) \tag{13.4}$$

$$\int_a^b f(t)dt + \int_b^c f(t)dt = \int_a^c f(t)dt \quad (\text{where } a \leq b \leq c) \tag{13.5}$$

Let $h: [a,b] \to R$ be defined by $h(s) := \int_a^s f(t)dt$. Then $h' \equiv f$ (where $a \leq b$) (13.6)

Let $\varphi: [a,b] \to [a_1,b_1]$ have $\varphi(a) = a_1$, $\varphi(b) = b_1$ (where $a \leq b$ and $a_1 \leq b_1$). Then, for $f: [a_1,b_1] \to R$,

$$\int_{a_1}^{b_1} f(t)dt = \int_a^b f(\varphi(s)) \cdot \varphi'(s)ds. \tag{13.7}$$

EXERCISES

13.1 (Hadamard). Assume Axiom 1 and the Integration Axiom. Prove that

$$\begin{aligned} &\forall f: R \to R \quad \exists g: R \times R \to R \text{ with} \\ &\forall (x,y) \in R \times R: f(x) - f(y) = (x-y) \cdot g(x,y) \end{aligned} \tag{13.8}$$

13.2 (Reyes). Prove that the g considered in (13.8) is unique provided we have the following axiom

$$\forall h: R \to R: (\forall x \in R: x \cdot h(x) = 0) \Rightarrow (h \equiv 0) \tag{13.9}$$

Hint: to prove uniqueness of g, it suffices to prove

$$(x-y) \cdot g(x,y) \equiv 0 \Rightarrow g(x,y) \equiv 0;$$

if $(x-y) \cdot g(x,y) \equiv 0$, substitute $z+y$ for x, to get $z \cdot g(z+y,y) \equiv 0$. Deduce from (13.9) that $g(z+y,y) \equiv 0$.

For models of (13.8) and (13.9), cf. III.Exercise 9.4.

13.3 (Reyes). Axxume Axiom 1, (13.8) and (13.9). Prove that $f'(x) = g(x,x)$, where f and g are related as in Exercise 13.1; prove also (without using integration) that

$$f(y) - f(x) = (y-x) \cdot f'(x) + (y-x)^2 \cdot h(x,y) \tag{13.10}$$

for some unique h. (Further iteration of Hadamard's lemma is also possible.) Prove that $h(x,x) = \frac{1}{2} f''(x)$.

13.4. Note that $\int_a^b f(t)dt$ is defined only when $a \leq b$. Why did we not have to make any assumptions of that kind in Proposition 13.1?

13.5. Prove $\int_{a_1}^{b_1} \int_{a_2}^{b_2} f(x_1,x_2)dx_2 dx_1 = \int_{a_2}^{b_2} \int_{a_1}^{b_1} f(x_1,x_2)dx_1 dx_2$.

I.14: FORMS AND CURRENTS

There are several ways of introducing the notion and calculus of differential forms in the synthetic context; for many objects, they will be equivalent. One way is a direct translation of the 'classical' one, others are related to form notions occurring in modern algebraic geometry. The various notions also have varying degree of generality in so far as the value object is conerned.

Let M be an arbitrary object, and V an object on which the multiplicative monoid $(R,\cdot)$ acts. Let $n \geq 0$ be a natural number. The following form notion is the one that (for $V = R$) mimicks the classical notion.

Definition 14.1. A differential n-form ω on M with values in V is a law which to any n-tuple $(t_1,\dots,t_n)$ of tangents to M with common base point associates an element $\omega(t_1,\dots,t_n) \in V$, in such a way that for any $\lambda \in R$ and $i = 1,\dots,n$,

$$\omega(t_1,\dots,\lambda\cdot t_i,\dots,t_n) = \lambda\cdot\omega(t_1,\dots,t_i,\dots,t_n) \tag{14.1}$$

and such that for any permutation σ of $\{1,\dots,n\}$, we have

$$\omega(t_{\sigma(1)},\dots,t_{\sigma(n)}) = \mathrm{sign}(\sigma)\cdot\omega(t_1,\dots,t_n) \tag{14.2}$$

In case V is an R-module satisfying the vector form of Axiom 1, it follows from Proposition 10.2 that (14.1) is actually the expected multilinearity condition, in cases where each tangent space T_mM is an R-module, in particular when M is infinitesimally linear. Hence we may refer to (14.1) as 'multilinearity'.

Also, (14.2) says that ω is alternating.

So an n-form on M with values in V is a fibrewise multilinear alternating map

$$TM \times_M TM \times_M \ldots \times_M TM \longrightarrow V$$

where the left hand side as usual denotes the 'n-fold pull-back':

$$[\![(t_1,\ldots,t_n) \in TM \times \ldots \times TM \mid t_i(0) = t_j(0) \quad \forall i,j]\!] .$$

The <u>object</u> of n-forms on M with values in V is then a subobject of

$$V^{TM \times_M \ldots \times_M TM}$$

carved out by certain finite inverse limit constructions.

Note that if M is infinitesimally linear, then

$$TM \times_M \ldots \times_M TM \cong M^{D(n)} ,$$

so that an n-form is a map

$$M^{D(n)} \longrightarrow V \tag{14.3}$$

satisfying certain conditions. The object of n-forms on M with values in V is thus a subobject of

$$V^{(M^{D(n)})}$$

Note that 0-forms are just functions $M \to V$.

We shall, however, mainly consider another form-notion. Let M be an arbitrary object, and $n \geq 0$ an integer. A map

$$D^n \xrightarrow{\tau} M$$

will be called an n-<u>tangent</u> at M. The object of these is M^{D^n}.

It carries n different actions of the multiplicative monoid $(R,\cdot)$, denoted γ_i:

$$\gamma_i : M^{D^n} \times R \longrightarrow M^{D^n} \qquad (i = 1,\dots,n)$$

where, for $\lambda \in R$ and $\tau \in M^{D^n}$, $\gamma_i(\tau,\lambda)$ is the composite of τ with the map $\lambda\cdot_i : D^n \to D^n$ which multiplies on the i'th coordinate by the scalar λ, and leaves the other coordinates unchanged.

Let V be an object with an action of the multiplicative monoid $(R,\cdot)$. The form-notion we now present is not always, but often (for many objects M), equivalent to the one given in Definition 14.1. For the rest of this §, we shall be dealing with this new notion.

<u>Definition 14.2</u>. A <u>differential</u> n-<u>form</u> ω on M with values in V is a map

$$M^{D^n} \xrightarrow{\ \omega\ } V$$

such that, for each $i = 1,\dots,n$,

$$\omega(\gamma_i(\tau,\lambda)) = \lambda\cdot\omega(\tau) \qquad \forall\, \tau \in M^{D^n}, \quad \forall\, \lambda \in R, \tag{14.4}$$

and such that for any permutation σ of $\{1,\dots,n\}$, we have

$$\omega(\tau \circ D^{\sigma}) = \mathrm{sign}(\sigma)\cdot\omega(\tau) \tag{14.5}$$

(where $D^{\sigma} : D^n \longrightarrow D^n$ permutes the n coordinates by σ).

Note that if M is infinitesimally linear, the inclusion $i: D(n) \subseteq D^n$ induces a restriction map $M^{D^n} \to M^{D(n)}$. Any differential form ω in the sense of Definition 14.1 gives rise (by viewing it as a map (14.3) and composing it with the restriction map) to a differential form $\widetilde{\omega}$ in the sense of Definition 14.2: for τ an n-tangent

$$\tilde{\omega}(\tau) := \omega(\tau \circ i) = \omega(\tau \circ \mathrm{incl}_1, \ldots, \tau \circ \mathrm{incl}_n)$$

where $\mathrm{incl}_i : D \to D^n$ injects D as the i'th axis.

The object of n-forms will be a subobject of $V^{(M^{D^n})}$, carved out by certain finite-inverse-limit constructions, corresponding to the equational conditions (14.4) and (14.5). We introduce the notation $E^n(M,V)$ for it, like in [17] p.355; $E^n(M,R)$ is just denoted $E^n(M)$. It is in an evident way an R-module. Note that $E^o(M,V) = V^M$.

The object $V^{M^{D^n}}$ itself will be considered later (§20) under the name: the object of infinitesimal (singular, cubical) n-cochains on M with values in V. A differential n-form in the sense of Definition 14.2 is such a cochain (with special properties).

A map $f: M \to N$ gives, by functorality, rise to a map $f^{D^n} : M^{D^n} \to N^{D^n}$, namely $\tau \longmapsto f \circ \tau$, for $\tau \in M^{D^n}$, and this map is compatible with the n different actions of $(R,\cdot)$ and of the permutation group in n letters. Therefore, if ω is a differential n-form on N, we get a differential n-form on M by composing with f^{D^n}; we denote it $f^*\omega$. We actually get a map

$$E^n(N,V) \xrightarrow{f^*} E^n(M,V).$$

In case V is an R-module, this f^* is R-linear.

Let V be an R-module

<u>Definition 14.3</u>. A (compact) n-<u>current</u> on M (relative to V) is an R-linear map

$$E^n(M,V) \longrightarrow V.$$

Thus, the object of n-currents on M (relative to V) is a subobject of $V^{E^n(M,V)}$, denoted $E_n(M,V)$.

The pairing

$$\mathcal{E}_n(M,V) \times \mathcal{E}^n(M,V) \longrightarrow V$$

will be denoted $\int$. Thus $\int_\gamma \omega = \gamma(\omega)$ if γ is an n-current and ω is an n-form. The contravariant functorality of the form notion gives immediately rise to covariant functorality of the current notion: if $f: M \to N$, and γ is an n-current on M, then $f_*\omega$ is the n-current on N given by

$$\int_{f_*\gamma} \omega := \int_\gamma f^*\omega$$

for any n-form ω on N.

Note that a 0-current (relative to R) is a distribution in the sense of Exercise 12.3.

Among the n-currents are some which we shall call 'infinitesimal singular n-rectangles'. They are given by pairs

$$(\tau,\underline{d}) \tag{14.6}$$

where $\tau: D^n \to M$ is an n-tangent, and $\underline{d} = (d_1,\ldots,d_n) \in D^n$. Such a pair gives rise to the n-current

$$\mathcal{E}^n(M,V) \longrightarrow V$$

given by

$$\omega \longmapsto d_1 \cdot \ldots \cdot d_n \cdot \omega(\tau) \tag{14.7}$$

It will be denoted $\langle\tau,\underline{d}\rangle$, so that in particular

$$\int_{\langle\tau,\underline{d}\rangle} \omega := d_1 \cdot \ldots \cdot d_n \cdot \omega(\tau) . \tag{14.8}$$

Let $\langle\tau,\underline{d}\rangle$ be such an infinitesimal singular n-rectangle on M. If $i = 1,\ldots,n$ and $\alpha = 0$ or 1, we define an infinitesimal singular (n-1)-rectangle on M,

$$F_{i,\alpha}\langle\tau,\underline{d}\rangle ,$$

"the $i\alpha$'th face of $\langle\tau,\underline{d}\rangle$", to be the pair consisting of the (n-1)-tangent

$$\tau(-,-,\ldots,\alpha\cdot d_i,\ldots,-)$$

and the (n-1)-tuple $(d_1,\ldots,\hat{d}_i,\ldots,d_n) \in D^{n-1}$ (d_i omitted).

We define the boundary $\partial\langle\tau,\underline{d}\rangle$ of $\langle\tau,\underline{d}\rangle$ to be the (n-1)-current

$$\sum_{i=1}^{n} \sum_{\alpha=0,1} (-1)^{i+\alpha} F_{i\alpha}\langle\tau,\underline{d}\rangle , \tag{14.9}$$

"the (signed) sum of the faces of the rectangle". The formula will be familiar from the singular cubical chain complex in algebraic topology.

We shall utilize the geometrically natural boundary (14.9) to define coboundaries of forms; for this, we shall assume that V is an R-module which satisfies Axiom 1. As a preliminary, we consider functions $\varphi: M^{D^n} \times D^n \to V$ which have the properties (for all $i = 1,\ldots,n$, all $\lambda \in R$, etc.):

$$\varphi(\gamma_i(\lambda,\tau),\underline{d}) = \lambda \cdot \varphi(\tau,\underline{d}) \tag{i}$$

$$\varphi(\tau,\lambda \cdot_i \underline{d}) = \lambda \cdot \varphi(\tau,\underline{d}) \tag{ii}$$

$$\varphi(\tau \circ D^\sigma,\underline{d}) = \operatorname{sign}(\sigma) \cdot \varphi(\tau,\underline{d}) . \tag{iii}$$

Clearly, the law (14.7) has these properties. But conversely

Proposition 14.4. Given $\varphi: M^{D^n} \times D^n \to V$ with properties (i), (ii), (iii), then there exists a unique differential n-form $\omega: M^{D^n} \to V$ with

$$\varphi(\tau,\underline{d}) = \int_{<\tau,\underline{d}>} \omega \quad (= d_1 \cdot \ldots \cdot d_n \cdot \omega(\tau)) \qquad \forall (\tau,\underline{d}) \in M^{D^n} \times D^n.$$

Proof. By (iii), $\varphi(\tau,\underline{d}) = 0$ if one of the coordinates of $\underline{d}$ is 0, so it is of form

$$\varphi(\tau,\underline{d}) = d_1 \cdot \ldots \cdot d_n \cdot \omega(\tau)$$

for some unique $\omega(\tau) \in V$, by "Property W_n" for V (Exercise 4.5). The fact that ω as a function of τ is multilinear and alternating (in the sense of (14.4), (14.5)) follows from (i) and (iii) above, together with the uniqueness assertion in W_n.

Let now θ be an (n-1)-form on M with values in V. The map $\varphi: M^{D^n} \times D^n \to V$ given by

$$(\tau,\underline{d}) \longmapsto \int_{\partial<\tau,\underline{d}>} \theta$$

is quite easily seen to have the properties (i), (ii), and (iii) in Proposition 14.4. We therefore have

Theorem 14.5. Given an (n-1)-form θ on M (with values in V). Then there exists a unique n-form (with values in V), denoted $d\theta$ (the 'coboundary of θ') so that for any current γ of form $<\tau,\underline{d}>$ with $\tau \in M^{D^n}$, $\underline{d} \in D^n$,

$$\int_{\partial\gamma} \theta = \int_{\gamma} d\theta . \tag{14.10}$$

We have not yet defined the boundary $\partial\gamma$ for arbitrary currents γ, only for those of the form $<\tau,\underline{d}>$. But now, of course, we may use (14.10) to define $\partial\gamma$ for any current γ. This $\partial\gamma$ will be considered in the next §.

Let us finally analyze more explicitely the 1-form df derived from a 0-form (= a function) $f: M \to V$. For an infinitesimal 1-rectangle $\langle\tau,d\rangle$, where $\tau: D \to M$, we have by (14.10)

$$\int_{\langle\tau,d\rangle} df = \int_{\partial\langle\tau,d\rangle} f$$

$$= f(\tau(d)) - f(\tau(0))$$

$$= d \cdot (f \circ \tau)'(0) \ . \tag{14.11}$$

Since also

$$\int_{\langle\tau,d\rangle} df = d \cdot df(\tau) \tag{14.12}$$

and (14.11) and (14.12) hold for all $d \in D$, we conclude, by cancelling the universally quantified d's, that

$$(df)(\tau) = (f \circ \tau)'(0), \tag{14.13}$$

the principal part of $f \circ \tau$. So df itself is the composite

$$df = M^D \xrightarrow{f^D} V^D \xrightarrow{\gamma} V \tag{14.14}$$

where γ is principal-part formation as in (7.4).

<u>EXERCISES</u>

14.1. Prove that the γ occurring in (7.4) and (14.14) may be viewed as the coboundary of the identity map $V \to V$ (which may itself be viewed as a V-valued 0-form on V).

Thus, the fact that for arbitrary $f: M \to V$, we have $df = \gamma \circ f^D$ can be deduced from naturality of the coboundary operator d:

$$df = d(id_V \circ f) = d(f^*(id_V)) = f^*(d(id_V)) = f^*(\gamma) = \gamma \circ f^D \ .$$

14.2. Let M be an R-module satisfying the vector form of Axiom 1. Identify M^{D^2} with M^4, via the map $M^4 \to M^{D^2}$ given by $(a,b_1,b_2,c) \longmapsto [(d_1,d_2) \longmapsto a + d_1 \cdot b_1 + d_2 \cdot b_2 + d_1 \cdot d_2 \cdot c]$. So a differential 2-form ω on M gets identified with a map $M^4 \to V$. The bilinearity of ω then implies that ω, for fixed a,b_1, depends linearily on (b_2,c), and for fixed (a,b_2), depends linearily on (b_1,c). So

$$\omega(a,b_1,b_2,c) = \omega(a_1,b_1,b_2,0) + \omega(a,b_1,0,c)$$

$$= \omega(a,b_1,b_2,0) + \omega(a,b_1,0,0) + \omega(a,0,0,c)$$

The second term vanishes. Hint: $\omega(a,b_1,b_2,0)$ depends bilinearily on b_1,b_2. The third term vanishes. Hint: ω is alternating.

This exercise contains the technique for proving equivalence of the form notions of Definitions 14.1 and 14.2 for suitable objects M.

I.15: CURRENTS DEFINED USING INTEGRATION. STOKES' THEOREM.

In this §, we shall assume a preorder relation $\leq$ on R, and the Integration Axiom of §13 (plus, of course, Axiom 1). We shall find it convenient to write $|[a,b]|$ for the set $[\![x \in R \mid a \leq x \leq b]\!]$ rather than $[a,b]$, as in §13. The notation $[a,b]$ will denote certain currents ("intervals") closely related to $|[a,b]|$, but with the information of the end points a and b retained.

Any R-valued n-form ω on a subset $U \subseteq R^n$, stable under addition of nilpotents in all n directions, determines a function $f: U \to R$, namely the unique one which satisfies

$$\omega((d_1,\ldots,d_n) \longmapsto (x_1+d_1,\ldots,x_n+d_n)) = d_1 \cdot \ldots d_n \cdot f(x_1,\ldots,x_n) \tag{15.1}$$

$\forall (x_1,\ldots,x_n) \in U$. It can be proved (see [45]) that the function f determines ω completely; so let us write $f\, dx_1 \cdots dx_n$ for ω.

Given an n-tuple of pairs $a_1 \leq b_1,\ldots,a_n \leq b_n$, we define a "canonical" n-current, denoted

$$[a_1,b_1] \times \ldots \times [a_n,b_n] \tag{15.2}$$

on the set $|[a_1,b_1]| \times \ldots \times |[a_n,b_n]| \subseteq R^n$, by putting

$$\int_{[a_1,b_1]\times\ldots\times[a_n,b_n]} \omega := \int_{a_n}^{b_n} \ldots \int_{a_1}^{b_1} f(x_1,\ldots,x_n)dx_1\ldots dx_n \, .$$

From (13.5) follows an 'additivity rule' for currents of form (15.2); e.g. for $a_1 \leq c_1 \leq b_1$ that the current (15.2) equals

$$[a_1,c_1]\times[a_2,b_2]\times\ldots\times[a_n,b_n]+[c_1,b_1]\times[a_2,b_2]\times\ldots\times[a_n,b_n]. \quad (15.3)$$

The n-current (15.2) has $2\cdot n$ (n-1)-currents as 'faces', defined much in analogy with §14. Specifically, the $i\alpha$'th face $(i=1,\ldots,n, \alpha=0,1)$ is obtained as

$$g_*([a_1,b_1]\times\ldots\times\underbrace{[a_i,b_i]}_{\text{omitted}}\times\ldots\times[a_n,b_n])$$

where

$$g(x_1,\ldots,x_{i-1},x_{i+1},\ldots,x_n) = (x_1,\ldots,x_{i-1},a_i,x_{i+1},\ldots,x_n)$$

if $\alpha = 0$, and similarly with b_i instead of a_i if $\alpha = 1$. A suitable alternating sum (in analogy with (14.9)) of these $2\cdot n$ (n-1)-currents is the 'geometric' boundary of the current (15.2).

Theorem 15.1 (Stokes). The geometric boundary $\bar{\partial}$ of the current (15.2) agrees with its current-theoretic boundary ∂ (recall that the latter was defined in terms of coboundary of forms).

Proof. We shall do the case $n = 2$ only. We first consider the case $b_2 = a_2+d_2$ (with $d_2 \in D$). Let θ be any (n-1)-form, i.e. a 1-form, on $|[a_1,b_1]| \times |[a_2,b_2]|$. We consider two functions g and h: $|[a_1,b_1]| \to R$, given by, respectively

$$g(c_1) = \int_{\bar{\partial}([a_1,c_1]\times[a_2,b_2])} \theta$$

$$h(c_1) = \int_{\partial([a_1,c_1]\times[a_2,b_2])} \theta = \int_{[a_1,c_1]\times[a_2,b_2]} d\theta \quad .$$

Clearly

$$g(a_1) = h(a_1) = 0 \tag{15.4}$$

We claim that, furthermore, $g' \equiv h'$. From the additivity rule (15.3) it is easy to infer

$$g(c_1+d_1) - g(c_1) = \int_{\partial([c_1,c_1+d_1]\times[a_2,a_2+d_2])} \theta,$$

(recalling $b_2 = a_2+d_2$), as well as

$$h(c_1+d_1) - h(c_1) = \int_{[c_1,c_1+d_1]\times[a_2,a_2+d_2]} d\theta .$$

But now we may note that we have an equality of currents

$$[c_1,c_1+d_1] \times [a_2,a_2+d_2] = <\tau,(d_1,d_2)>,$$

where $\tau\colon D \times D \to R \times R$ is just 'parallel transport to (c_1,a_2)', (i.e. $(\delta_1,\delta_2) \longmapsto (c_1+\delta_1,a_2+\delta_2)$); for, they take on a 2-form $f(x,y)dxdy$ value, respectively

$$\int_{c_1}^{c_1+d_1} \int_{a_2}^{a_2+d_2} f(x,y)dxdy$$

and

$$d_1 \cdot d_2 \cdot f(c_1,a_2)$$

by (15.1); these two expressions agree, by twofold application of "the fundamental theorem of calculus" (13.6). We conclude that $g'(c_1) = h'(c_1)$, and from the uniqueness assertion in the integration axiom and (15.4) we conclude $g \equiv h$, in particular $g(b_1) = h(b_1)$. This proves Stokes' Theorem for "long thin" rectangles $[a_1,b_1] \times [a_2,a_2+d_2]$. The passage from these to arbitrary rectangles proceeds similarly by the uniqueness in the integration axiom, now using the result proved for the long thin rectangles to to deduce equality of the respective derivatives.

In the following, I^n denotes both $|[0,1]| \times \ldots \times |[0,1]| \subseteq R^n$ as well as the n-current $[0,1] \times \ldots \times [0,1]$.

Let ω be an n-form on M, and let $f: I^n \to M$ be an arbitrary map (a "singular n-cube in M"). We may define an n-current (also denoted f) on M by putting

$$\int_f \omega := \int_{I^n} f^*\omega$$

for ω an n-form on M; - equivalently, $f = f_*(I^n)$. The geometric boundary $\bar{\partial}f$ of f is defined by

$$\int_{\bar{\partial}f} \theta := \int_{\bar{\partial}(I^n)} f^*\theta,$$

or equivalently $\bar{\partial}f = f_*(\bar{\partial}(I^n))$. It is a sum of 2n (n-1)-currents of form $I^{n-1} \to M$.

Corollary 15.2. Let θ be an (n-1)-form on M, and $f: I^n \to M$ a map. Then

$$\int_f d\theta = \int_{\bar{\partial}f} \theta .$$

Proof. We have

$$\int_f d\theta = \int_{f_*(I^n)} d\theta = \int_{\partial(f_*(I^n))} \theta$$

(by definition of boundary of currents)

$$= \int_{f_*(\partial(I^n))} = \int_{\partial(I^n)} f^*\theta$$

$$= \int_{\bar{\partial}(I^n)} f^*\theta,$$

the last equality by the theorem.

In summary: we first defined the <u>geometric</u> boundary for infinitesimal currents, and then <u>defined</u> coboundary of forms in terms of that, in other words, so as to make Stokes' Theorem true-by-definition for infinitesimal currents. Then we defined boundary of arbitrary currents in terms of coboundary of forms. So Stokes' Theorem for the current-theoretic boundary is again tautological. The non-trivial Stokes' Theorem then consists in proving that the current-theoretic boundary agrees with the geometric boundary, and this comes about by reduction to the infinitesimal case where it was true by construction.

I.16: WEIL ALGEBRAS

Let k be a commutative ring in the category Set. In the applications, k will be $\mathbb{Z}$, $\mathbb{Q}$, or $\mathbb{R}$. A Weil algebra structure on k^n is a k-bilinear map

$$\mu: k^n \times k^n \longrightarrow k^n$$

making k^n (with its evident k-module structure) into a commutative k-algebra with $(1,0,\ldots,0)$ as multiplicative unit, and such that the set I of elements in k^n with first coordinate zero is an ideal and has $I^n = 0$ (meaning: the product under μ of any n elements from I is zero). In particular, each element of form $(0,x_2,\ldots,x_n)$ is nilpotent.

A Weil algebra over k is a k-algebra W of form (k^n,μ) with μ a Weil algebra structure on k^n. Each Weil algebra comes equipped with a k-algebra map

$$\pi: W \longrightarrow k$$

given by $(x_1,\ldots,x_n) \longmapsto x_1$, called the augmentation. Its kernel is I.

The basic examples of Weil algebras are k itself and $k[\varepsilon] = k \times k$.

Since the map μ is k-bilinear, it is described by an $n \times n^2$ matrix $\{\gamma_{ijk}\}$ with entries from k, namely

$$\mu(\underline{e}_j,\underline{e}_k) = \sum_{i=1}^{n} \gamma_{ijk}\, \underline{e}_i$$

where $\underline{e}_i = (0,\ldots,0,1,0,\ldots,0)$ (1 in the i'th position). Equivalently

$$\mu((x_1,\ldots,x_n),(y_1,\ldots,y_n)) =$$

$$(\sum_{jk} \gamma_{1jk}\, x_j y_k, \ldots, \sum_{jk} \gamma_{njk}\, x_j y_k). \tag{16.1}$$

The condition $I^n = 0$ is a purely equational condition on the 'structure constants' γ_{ijk}.

Suppose now that R is a commutative k-algebra object in a category E with finite products. Then the description (16.1) defines an R-bilinear map $\mu_R: R^n \times R^n \longrightarrow R^n$ making R^n into a commutative R-algebra object with $(1,0,\ldots,0)$ as multiplicative unit; we denote it $R \otimes W$,

$$R \otimes W := (R^n, \mu_R).$$

There is a canonical R-algebra map π, the augmentation, namely projection to 1'st factor. Its kernel is canonically isomorphic to R^{n-1}, and is denoted $R \otimes I$ ('the augmentation ideal'). The composite

$$(R \otimes I)^n \rightarrowtail (R \otimes W)^n \xrightarrow{\bar{\mu}_R} R \otimes W \tag{16.2}$$

where $\bar{\mu}_R$ is iterated multiplication, is the zero map, because it is described entirely by a certain combination of the structure constants which is zero by the assumption $I^n = 0$.

Each Weil algebra $W = (k^n, \mu)$ is a finitely presented k-algebra (n generators will suffice; sometimes fewer will do, like for $k[\varepsilon]$ where one generator ε suffices).

If R is a k-algebra object in a category with finite inverse limits, objects of form $\mathrm{Spec}_R(W)$, for some Weil algebra over k, are called <u>infinitesimal</u> objects (relative to R), (or formal-infinitesimal objects, more precisely). Each such has a canonical base point b

$$\mathbf{1} \xrightarrow{b=\mathrm{Spec}_R(\pi)} \mathrm{Spec}_R(W)$$

induced by the augmentation $W \to k$ (note $\mathrm{Spec}_R(k) = \mathbf{1}$). If $W = k[\varepsilon]$, $\mathrm{Spec}_R(W) = D$, and the base point is

$$\mathbf{1} \xrightarrow{0} D = \mathrm{Spec}_R(k[\varepsilon]).$$

Of course $\mathbf{1}$, being $\mathrm{Spec}_R(k)$, is also infinitesimal.

If E is furthermore cartesian closed, we shall prove, for any R-algebra C

$$\underset{\sim}{\mathrm{hom}}_{R\text{-Alg}}(R \otimes W, C) \cong \mathrm{Spec}_C(W), \tag{16.3}$$

in a way which is natural in C. For,

$$\underset{\sim}{\mathrm{hom}}_{R\text{-Alg}}(R \otimes W, C) \subseteq \underset{\sim}{\mathrm{hom}}_{R\text{-lin}}(R^n, C) \cong C^n$$

and the subobject here is the extension of the formula "multiplication is preserved", i.e. is the sub"set"

$$[\![(c_1, \ldots, c_n) \in C^n \mid c_j c_k = \Sigma\, \gamma_{ijk}\, c_i \quad \forall j,k]\!]$$

(using set-theoretic notation for the equalizer in question), where the γ_{ijk} are the structure constants. This object, however, is also $\mathrm{Spec}_C(W)$, as is seen by choosing the following presentation of W:

$$k[X_1, \ldots, X_n] \longrightarrow W = k^n$$

with $X_i \longmapsto \underline{e}_i$ and with kernel the ideal generated by

$$X_j X_k - \sum_i \gamma_{ijk} X_i \qquad \forall j,k$$

If R furthermore satisfies Axiom 2_k, we have also the isomorphism ν:

$$\underset{\sim\sim}{\mathrm{hom}}_{R\text{-Alg}}(R^{\mathrm{Spec}_R(W)}, C) \cong \mathrm{Spec}_C(W),$$

whence we (by Yoneda's lemma) get an isomorphism

$$\alpha\colon R \otimes W \cong R^{\mathrm{Spec}_R(W)} .$$

The isomorphism α here is a straightforward generalization of the $\alpha\colon\ R[\varepsilon] \to R^D$ of Axiom 1. In fact, the exponential adjoint $^{\vee}\alpha$ of α makes the triangle

$$\begin{array}{ccc}
(R \otimes W) \times \mathrm{Spec}_R(W) & \xrightarrow{\ {}^{\vee}\alpha\ } & R \\
\cong \big\uparrow & \nearrow_{\mathrm{ev}} & \\
(R \otimes W) \times \underset{\sim\sim}{\mathrm{hom}}_{R\text{-Alg}}(R \otimes W, R) & &
\end{array} \qquad (16.4)$$

commutative, where the vertical isomorphism is derived from (16.3) (with $C = R$), and ev denotes evaluation.

In the case where there is given an explicite presentation of W

$$p\colon k[X_1, \ldots, X_h] \twoheadrightarrow W = k^n$$

with kernel I, and if the polynomials $\varphi_j = \varphi_j(X_1, \ldots, X_h)$ $(j = 1, \ldots, n)$ have $p(\varphi_j) = \underline{e}_j$, then, with $D' = \mathrm{Spec}_R W \subseteq R^h$, the $^{\vee}\alpha$ may be described explicitely as

$$((t_1, \ldots, t_n), (d_1, \ldots, d_h)) \longmapsto \sum_{j=1}^{n} t_j \varphi_j(d_1, \ldots, d_h) \qquad (16.5)$$

We may summarize:

Theorem 16.1. Axiom 2_k for R implies Axiom 1_k^W for R, where we pose

Axiom 1_k^W. For any Weil algebra W over k, the R-algebra homomorphism

$$R \otimes W \xrightarrow{\alpha} R^{\mathrm{Spec}_R(W)}$$

is an isomorphism (where α is the exponential adjoint of the $^{\vee}\alpha$ described in (16.4)).

The Axiom implies Axiom 1 (take $W = k[\varepsilon]$), Axiom 1' (take $W = k[X]/(X^{k+1})$, and even Axiom 1" (see Exercise 1).

Using the explicit description (16.5) of $^{\vee}\alpha$, and with W and D' as there, Axiom 1_k^W, for this Weil algebra, may be given the naive verbal form (where the φ_j's are certain fixed polynomials with coefficients from k):

"Every map $f: D' \to R$ is of form

$$(d_1,\ldots,d_h) \longmapsto \sum_{j=1}^{n} t_j \cdot \varphi_j(d_1,\ldots,d_h) \quad \forall (d_1,\ldots,d_h) \in D'$$

for unique $t_1,\ldots,t_n \in R$. "

Proposition 16.2. Axiom 1_k^W for R implies that R is infinitesimally linear, has Property W, and, if k contains $\mathbb{Q}$, the Symmetric Functions Property (4.5).

Proof. Property W follows from Axiom 1 alone, as noted in Exercise 4.2, and the Symmetric Functions Property follows from Axiom 1', by Exercise 4.4 (provided $\mathbb{Q} \subseteq k$). Finally, infinitesimal linearity follows from Axiom 1", by Proposition 6.4.

Proposition 16.3. Axiom 1_k^W for R implies that any map

$$\mathrm{Spec}_R(W) \to R,$$

taking the base point b to 0, has only nilpotent values, i.e. factors through some D_k.

Proof. Under the identification $R\otimes W \cong R^{\mathrm{Spec}_R W}$, the maps $\mathrm{Spec}_R W \to R$ with $b \mapsto 0$ correspond to the elements $\underline{r} \in R\otimes W$ with $\pi(\underline{r}) = 0$, i.e. to the elements of the augmentation ideal $R\otimes I$. But such elements are nilpotent, since the composite (16.2) is zero.

(The Proposition is also true in 'parametrized' form: for any object X, and any map $g: X \times \mathrm{Spec}_R(W) \to R$ such that the composite

$$X \xrightarrow{\langle X,b\rangle} X \times \mathrm{Spec}_R(W) \xrightarrow{\ g\ } R$$

is constant 0, g has only nilpotent values.)

We finish this § by describing a class of Weil algebras that will be used in §18, and whose spectra will be denoted $\tilde{D}(p,q) \subseteq R^{p\cdot q}$, $(p\leq q)$. We assume that k is a $\mathbb{Q}$-algebra. Let $W(p,q)$ be the Weil algebra given by the presentation (i ranging from 1 to p, j from 1 to q)

$$W(p,q) = k[X_{ij}]/(X_{ij}\cdot X_{i'j'} + X_{ij'}\cdot X_{i'j})$$

Note that since 2 is invertible, we may deduce that

$$X_{ij}\cdot X_{ij'} = 0 \quad \text{in} \quad W(p,q) \tag{16.6}$$

Theorem 16.4. A k-linear basis for $W(p,q)$ is given by those polynomials that occur as minors (= subdeterminants) of the $p\times q$ matrix of indeterminates $\{X_{ij}\}$ (including the "empty" minor, which is taken to be 1).

A proof will be given in Exercise 16.4 below.

Because of (16.6), the $\tilde{D}(p,q)$ (= $\mathrm{Spec}_R W(p,q)$) is contained in $D(q) \times \ldots \times D(q) \subseteq R^{p\cdot q}$ (p copies of $D(q)$).

Axiom 1^W, for $W(p,q)$, expresses, in view of the explicit linear basis given above, that any map $\tilde{D}(p,q) \to R$ is given by a linear combination, with uniquely determined coefficients from R, of the 'subdeterminant' functions $R^{p\cdot q} \to R$. In particular, for $p = 2$, it is of form, for unique $a, \alpha_j \beta_j$, and $\gamma_{jj'}$,

$$\begin{Bmatrix} d_{11} \cdots d_{1q} \\ d_{21} \cdots d_{2q} \end{Bmatrix} \longmapsto a + \alpha_j d_{1j} + \beta_j d_{2j} + \sum_{j<j'} \gamma_{jj'} \cdot \begin{vmatrix} d_{1j} & d_{1j'} \\ d_{2j} & d_{2j'} \end{vmatrix},$$

so that a function $\tilde{D}(2,n) \to R$ which vanishes on the two "copies of $D(n)$", $D(n) \times \{0\}$ and $\{0\} \times D(n)$ in $\tilde{D}(2,n)$, is of form

$$\begin{Bmatrix} d_{11} \cdots d_{1n} \\ d_{21} \cdots d_{2n} \end{Bmatrix} \longmapsto \sum_{j<j'} \gamma_{jj'} \cdot \begin{vmatrix} d_{1j} & d_{1j'} \\ d_{2j} & d_{2j'} \end{vmatrix}$$

with unique $\gamma_{jj'}$. More generally, a function $\tilde{D}(p,n) \to R^m$ vanishing on the p copies of $\tilde{D}(p-1,n)$ is of form

$$\underline{\underline{D}} \longmapsto \sum_L \gamma_L \cdot (\text{L'th } p \times p \text{ minor of } \underline{\underline{D}})$$

where L ranges over the set $\binom{n}{p}$ of $p \times p$ minors of $\underline{\underline{D}}$. The subobject of such functions is denoted $[\tilde{D}(p,n), R^m]$; thus

$$[\tilde{D}(p,n), R^m] = \underset{\sim}{\hom}_{\text{p-lin;alternating}}(\underbrace{R^n \times \ldots \times R^n}_{p}, R^m) \tag{16.7}$$

The geometric significance of $\tilde{D}(p,n) \subseteq D(n) \times \ldots \times D(n)$ (p copies) is the following

<u>Proposition 16.5</u>. Let $\underline{d}_i \in D(n)$ $(i = 1,\ldots,p)$. Then

$$(\underline{d}_1,\ldots,\underline{d}_p) \in \widetilde{D}(p,n)$$

iff

$$\underline{d}_i - \underline{d}_{i'} \in D(n) \qquad \forall i,i' = 1,\ldots,p.$$

Proof. The latter condition expresses (writing $\underline{d}_i = (d_{i1},\ldots,d_{in})$) that for any $j,j' = 1,\ldots,n$

$$(d_{ij} - d_{i'j})\cdot(d_{ij'} - d_{i'j'}) = 0$$

which, in view of $d_{ij}\, d_{ij'} = 0$ (because $\underline{d}_i \in D(n)$), and similarly for i', means

$$-d_{ij}\, d_{i'j'} - d_{i'j}\, d_{ij'} = 0$$

which are exactly (minus one times) the equations definining $\widetilde{D}(p,n)$.

EXERCISES

16.1. Describe Weil algebras (over $\mathbb{Z}$) whose Spec_R are the infinitesimal objects considered in §6:

$$D_k(n),\ \widetilde{D}(2,n),\ D_c\ .$$

Using the Weil algebra for $D_k(n)$, prove that Axiom $1^W_{\mathbb{Z}}$ implies Axiom 1".

16.2. Prove that if W_1 and W_2 are Weil algebras over k, then so is $W_1 \otimes_k W_2$. Conclude that the product of two infinitesimal objects is infinitesimal.

16.3. Let k be a field of characteristic 0. Make $k[X_1,\ldots,X_n]$ into a module over $k[\frac{\partial}{\partial x_1},\ldots,\frac{\partial}{\partial x_n}]$ by letting $\frac{\partial}{\partial x_i}$ act on polynomials by partial differentiation. Prove that any finitely generated submodule $E \subseteq k[X_1,\ldots,X_n]$ is finite dimensional as a vector space.

According to Emsalem [16], if $\varphi_1,\ldots,\varphi_m$ is a k-basis for such E, the k-linear dual E^* of E has a k-basis the functionals

$$Q \longmapsto \varphi_i(\frac{\partial}{\partial X_1},\ldots,\frac{\partial}{\partial X_n})(Q)(0) \qquad i = 1,\ldots,m$$

(where $Q \in k[X_1,\ldots,X_n]$). Verify this for the case of one variable, i.e. for $n = 1$.

Prove (Emsalem) that if $J \subseteq k[\frac{\partial}{\partial X_1},\ldots,\frac{\partial}{\partial X_n}]$ is the ideal of those elements whose action on E is zero, then $k[\frac{\partial}{\partial X_1},\ldots,\frac{\partial}{\partial X_n}]/J$ is a Weil algebra W of k-linear dimension m.

So as an algebra, W can be interpreted as an algebra of differential operators $E \to E$.

Prove that if $n = 1$ and E is the $k[\frac{\partial}{\partial X}]$-submodule generated by X^m, the resulting W is the one defininining D_m.

Similarly, if $n = 2$, and E is the submodule generated by $X_1 \cdot X_2$, we get $D \times D$. Also $X_1^2 + X_2^2$ yields the D_c of (6.6).

Finally, if $n = 2$, and E is the submodule generated by X_1 and X_2, we get $D(2)$.

16.4. (I am indebted to H.A. Nielsen for providing the following proof of Theorem 16.4.). We use multilinear algebra as described in, say [47], Chapter XVI. Let k be a commutative $\mathbb{Q}$-algebra (in Sets); $\otimes$ denotes $\otimes_k$. Let E and F denote the k-modules k^p and k^q, respectively.

Identify the ring $k[X_{11},\ldots,X_{pq}]$ in a $p \cdot q$ matrix of indeterminates with the symmetric k-algebra $S^{\cdot}(E \otimes F)$, and the Weil algebra $W(p,q)$, considered above, with the quotient ring

$$S^{\cdot}(E \otimes F)/I$$

where I is the ideal generated by the image of the embedding

$$S^2E \otimes S^2F \longrightarrow S^{\cdot}(E \otimes F)$$

given by

$$(e_1 \otimes e_2) \otimes (f_1 \otimes f_2) \longmapsto (e_1 \otimes f_1)\cdot(e_2 \otimes f_2) + (e_1 \otimes f_2)(e_2 \otimes f_1).$$

Consider also the k-module L:

$$\bigoplus_r \Lambda^r(E) \otimes \Lambda^r(F),$$

and prove that it becomes a commutative k-algebra by putting

$$((e_1\wedge\ldots\wedge e_r) \otimes (f_1\wedge\ldots\wedge f_r))\cdot((e_1'\wedge\ldots\wedge e_s') \otimes (f_1'\wedge\ldots\wedge f_s'))$$

$$= (e_1\wedge\ldots\wedge e_r\wedge e_1'\wedge\ldots\wedge e_s') \otimes (f_1\wedge\ldots\wedge f_r\wedge f_1'\wedge\ldots\wedge f_s').$$

Define a k-algebra homomorphism $\psi\colon S^{\cdot}(E \otimes F) \to L$ by putting

$$\psi(e \otimes f) = e \otimes f \in \Lambda^1E \otimes \Lambda^1F,$$

and prove that it vanishes on the ideal I, defininining a k-algebra homomorphism $\bar{\psi}\colon S^{\cdot}(E \otimes F)/I \to L$.

Define a k-module map $\varphi\colon L \to S^{\cdot}(E \otimes F)$ by

$$\varphi((e_1 \wedge \ldots \wedge e_r) \otimes (f_1 \wedge \ldots \wedge f_r)) = \frac{1}{r!}\cdot\begin{vmatrix} e_1 \otimes f_1 & \ldots & e_1 \otimes f_r \\ \cdot & & \\ \cdot & & \\ e_r \otimes f_1 & \ldots & e_r \otimes f_r \end{vmatrix},$$

and prove that the composite

$$L \xrightarrow{\varphi} S^{\cdot}(E \otimes F) \longrightarrow S^{\cdot}(E \otimes F)/I$$

is a k-algebra homomorphism $\bar{\varphi}$.

Identifying $S^{\cdot}(E \otimes F)$ with $k[X_{11},\ldots,X_{pq}]$, prove that the canonical basis for $\Lambda^r E \otimes \Lambda^r F$ by φ goes into the set of $r \times r$ subdeterminants of the X_{ij}'s (except for an invertible scalar factor), and deduce the validity of Theorem 16.4.

I.17: FORMAL MANIFOLDS

We would like a class of geometric objects where one can define a k-neighbour relation $\sim_k$ generalizing that of §6 for the R^n, and on which one 'locally' can prove things by choosing local coordinate charts. For this purpose we introduce the notion of formal-étale inclusion, and later the stronger notion of open inclusion; out of these notions, one arrives at the notion of formal manifold, respectively manifold. On formal manifolds, and in particular on manifolds, the neighbour relation and local coordinate calculations make sense.

The notion of 'formal manifold' to be introduced here is slightly stronger than the one of [35], since we here shall consider external covering families. This is done in order to keep the categorical logic at a lower level.

If $\mathcal{D}$ is a class of maps in a cartesian closed category $\mathcal{E}$, we say that a map $f: M \to N$ is $\mathcal{D}$-étale if for each $j: J \to K$ in $\mathcal{D}$, the commutative square

$$\begin{array}{ccc} M^K & \xrightarrow{f^K} & N^K \\ {\scriptstyle M^j}\downarrow & & \downarrow{\scriptstyle N^j} \\ M^J & \xrightarrow[f^J]{} & N^J \end{array} \qquad (17.1)$$

is a pull-back. (If $\mathcal{D}$ is small in a suitable sense, and $\mathcal{E}$ is suitably left complete, we may even form the'object of $\mathcal{D}$-étale maps from M to N' as a certain subobject of N^M; this is needed for the approach of [35], but in the present set up, it suffices to be able to talk about the (external) class of $\mathcal{D}$-étale maps from M to N.)

If $f_1: M_1 \to N_1$ and $f_2: M_2 \to N_2$ are $\mathcal{D}$-étale, then using that $(-)^K$ and $(-)^J$ preserve products, it is easy to see that $f_1 \times f_2: M_1 \times M_2 \to N_1 \times N_2$ is likewise $\mathcal{D}$-étale. Similarly for several factors.

The composite of two $\mathcal{D}$-étale maps is $\mathcal{D}$-étale. Also, from a well-known box lemma about pull-back diagrams, and the fact that $(-)^J$ and $(-)^K$ preserve pull-backs, it follows that in a pull-back diagram

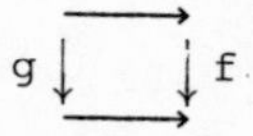

(17.2)

we have: f $\mathcal{D}$-étale ⇒ g $\mathcal{D}$-étale, ("$\mathcal{D}$-étaleness is stable under pull-back").

In set-theoretic terms, the condition that (17.1) is a pull-back says: given $g \in N^K$ and $h \in M^J$ so that $g \circ j = f \circ h$, there exists a unique $\ell \in M^K$ with

$$f \circ \ell = g \quad \text{and} \quad \ell \circ j = h \;;$$

diagrammatically, given a commutative square as below, there exists a unique diagonal fill-in (dotted arrow) making the two triangles commute:

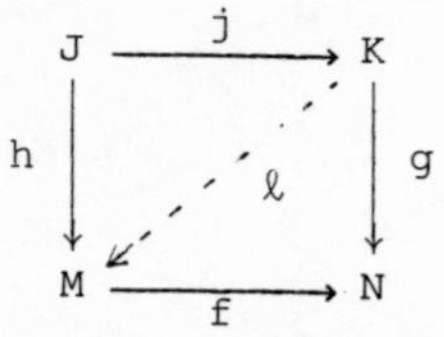

This description of $\mathcal{D}$-étaleness is adequate only in the naive approach. The full description without exponentials is that for any object X ("parameter object") and any $j \in \mathcal{D}$, any commutative square

$$\begin{array}{ccc} J\times X & \xrightarrow{j\times X} & K\times X \\ \downarrow & \swarrow & \downarrow \\ M & \xrightarrow[f]{} & N \end{array}$$

has a diagonal fill-in (dotted arrow).

For the rest of this §, R is a k-algebra object in a finitely complete cartesian closed category $\mathcal{E}$, and is assumed to satisfy Axiom 2 or at least Axiom 1^W; $\mathcal{D}$ is the class of base points $\mathbf{1} \longrightarrow \mathrm{Spec}_R W$ of infinitesimal objects, in the sense of §16. The $\mathcal{D}$-étale maps are then called <u>formal-étale</u>. We assume that $\mathcal{E}$ has good exactness properties, say, is a topos. In particular, we can form D_∞ as in §5.

<u>Proposition 17.1</u>. The inclusion $(D_\infty)^n \rightarrowtail R^n$ is formal-étale.

<u>Proof</u>. Since the product of $\mathcal{D}$-étale maps is $\mathcal{D}$-étale, it suffices to consider $D_\infty \rightarrowtail R$. Arguing naively, the result is then contained in Proposition 16.3. Arguing generally (with generalized elements, as in Part II), one needs the "parametrized form" of Proposition 16.3.

We now agree to say that an object U that appears as domain of a formal-étale monic map into R^n is an n-<u>dimensional model</u> object. Examples are D_∞^n (by Proposition 17.1), or R^n; also, for $n = 1$, the object $\mathrm{Inv}(R) \rightarrowtail R$ 'of invertible elements in R' will be a 1-dimensional model object, giving rise to many other model objects, as we shall see in §19.

A <u>formal n-dimensional manifold</u> is now defined to be an object M for which there exists a jointly epic class of monic formal-étale maps

$$U_i \rightarrowtail M \qquad i \in I$$

with each U_i an n-dimensional model.

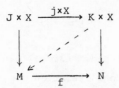

Clearly, any n-dimensional model U, and in particular R^n itself, is a formal n-dimensional manifold.

For each integer $k \geq 0$, and each formal n-dimensional manifold M, we shall introduce a binary 'k-neighbour' relation $\sim_k$, generalizing the one considered in §6 for R^n. As a subobject of $M \times M$, it will be denoted $M_{(k)} \subseteq M \times M$

$$M_{(k)} = [\![(x,y) \in M \times M \mid x \sim_k y]\!] .$$

It will be convenient to think in terms of the following completely general notions. Suppose M and N are objects in a category with finite inverse limits, and that $S_M \subseteq M \times M$ and $S_N \subseteq N \times N$ are binary relations on M and N, respectively. We say, then, that a map $\varphi: M \to N$

<u>preserves</u> the S-relation if $S_M \subseteq (\varphi \times \varphi)^{-1}(S_N)$

<u>reflects</u> the S-relation if $S_M = (\varphi \times \varphi)^{-1}(S_N)$

<u>creates</u> the S-relation if φ preserves the S-relation and the following square is a pull-back:

$$\begin{array}{ccc} S_M & \xrightarrow{\varphi \times \varphi} & S_N \\ {\scriptstyle p_1}\downarrow & & \downarrow{\scriptstyle p_1} \\ M & \xrightarrow[\varphi]{} & N. \end{array}$$

Here, p_1 denotes projection to the first factor. Clearly, if φ is monic and creates the S-relation, then it reflects it. In set theoretic terms

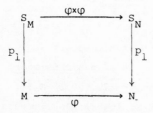

φ <u>preserves</u> S: $\forall\, x,y \in M \quad x\,S\,y \Rightarrow \varphi(x)\,S\,\varphi(y)$

φ <u>reflects</u> S: $\forall\, x,y \in M \quad x\,S\,y \Leftrightarrow \varphi(x)\,S\,\varphi(y)$

φ <u>creates</u> S: φ preserves S, and
$\forall x \in M \quad \forall z \in N$: $\varphi(x)\,S\,z \Rightarrow \exists!\; y \in M$ with $x\,S\,y$ and $\varphi(y) = z$.

Recall the binary relation $\sim_k$ on R^n, considered in §6. The subobject $(R^n)_{(k)} \subseteq R^n \times R^n$ may be identified with

$$R^n \times D_k(n) \xrightarrow{<proj_1,+>} R^n \times R^n .$$

<u>Lemma 17.2</u>. Assume $\varphi: U \rightarrowtail R^n$ is a formal-étale monic, and define the binary relation $\sim_k$ on U so that φ reflects it, i.e. $U_{(k)} = (\varphi\times\varphi)^{-1}(R^n{}_{(k)})$. Then φ creates the relation $\sim_k$.

<u>Proof</u>. Consider the commutative diagram

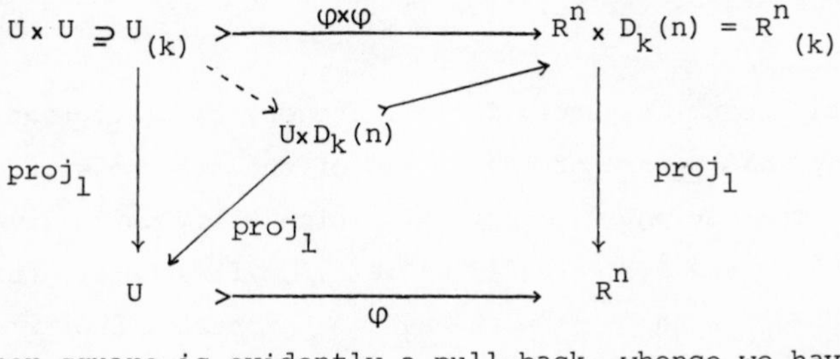

The inner square is evidently a pull-back, whence we have the (monic) comparison map (dotted arrow). We want to produce an inverse for it. Since $U_{(k)} \subseteq U \times U$, we should produce two maps $U \times D_k(n) \to U$. One of them is just $proj_1$. The other one we get by using the formal étaleness condition ("the parametrized form") for φ on the commutative diagram

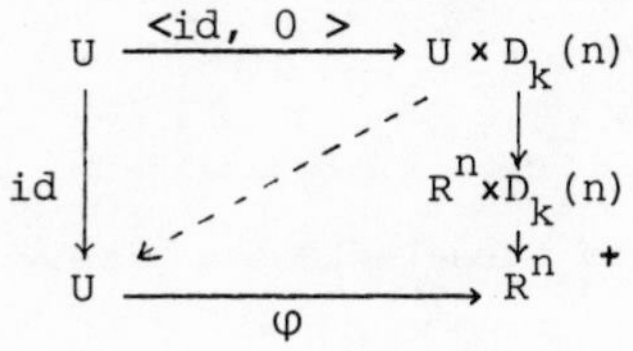

to produce the dotted arrow.

From the Lemma we deduce in particular that if we view U as a subobject of R^n via φ, in naive terms

$$\underline{x} \in U \wedge \underline{d} \in D_k(n) \Rightarrow \underline{x} + \underline{d} \in U.$$

Combining this with Corollary 6.5, it is easy to see that the relation $\sim_k$ on an n-dimensional model object U does not depend on the monic formal-étale $U \rightarrowtail R^n$ chosen, so it is canonical.

The following Lemma is now almost trivial

Lemma 17.3. If $V \rightarrowtail U$ is a monic formal-étale map and U is an n-dimensional model object, then so is V, and the map creates the relation $\sim_k$.

Proposition 17.4. Let M be a formal n-dimensional manifold. Then there exists precisely one binary relation $\sim_k$ on M such that for any formal-étale monic $\varphi: U \rightarrowtail M$ with U an n-dimensional model, φ creates the relation $\sim_k$.

Proof. This is now a purely formal category-theoretic consequence of the previous part of the §, and of the exactness properties assumed for the ambient category $\mathcal{E}$, which allow us to give the proof as if $\mathcal{E}$ were the category of sets. We first prove the uniqueness. Suppose $\sim$ and $\sim'$ both have the property. Let $x \sim y$. By assumption, there is a formal-étale monic $\varphi: U \rightarrowtail M$ from a model object U with $x \in \varphi(U)$, so $x = \varphi(\xi)$ with $\xi \in U$. Since φ creates the relation $\sim$, there exists an $\eta \in U$ with $\varphi(\eta) = y$ and $\xi \sim_k \eta$. Since φ takes the relation $\sim_k$ on U to the relation $\sim'$ on M, we conclude that $x \sim' y$.

To prove the existence, we define $M_{(k)} \subseteq M \times M$ to be

$$\bigcup_{i \in I} (\psi_i \times \psi_i)(U^i_{(k)})$$

where $\psi_i: U^i \rightarrowtail M$ is a covering class of formal-étale monics.

We claim that each ψ_j then creates $\sim_k$. For, suppose $x \sim y$ in M (writing everywhere $\sim$ for $\sim_k$). So there is a $\psi_i \colon U^i \to M$ with $\psi_i^{1}(x) \sim \psi_i^{-1}(y)$ in U_i. Let $\xi \in U^j$ have $\psi_j(\xi) = x$. Consider $U^i \times_M U^j$. The inclusion of this into U^i is formal-étale (formal-étale maps being stable under pull-back), so it creates $\sim$ by Lemma 17.3. So not only do we, by assumption have $\bar{\xi}$ there, going to $\psi_i^{-1}(x)$, but also an $\bar{\eta}$ going to $\psi_i^{-1}(y)$, and with $\bar{\xi} \sim \bar{\eta}$. Since the inclusion of $U^i \times_M U^j$ into U^j preserves $\sim$, the image of $\bar{\eta}$ in U^j witnesses that ψ_j creates $\sim$.

The argument that an arbitrary formal-étale monic $\varphi \colon U \to M$ creates $\sim_k$ is similar (U a model object).

In analogy with the notation of §6 we may, for any formal manifold M and any $x \in M$ write $\mathcal{M}_k(x)$ for "the k-monad around x",

$$\mathcal{M}_k(x) := [\![y \in M \mid x \sim_k y]\!]$$

which is useful when thinking naively, (cf. the remarks in §6 about these monads). For instance, saying that a monic map $f \colon M \to N$ creates $\sim_k$ can be expressed: f maps $\mathcal{M}_k(x)$ bijectively onto $\mathcal{M}_k(f(x))$. Also, $\mathcal{M}_k(x)$ is isomorphic to $D_k(m)$ (where $m = \dim(M)$).

Proposition 17.5. Let $f \colon M \to N$ be any map between formal manifolds (not necessarily of same dimension). Then f preserves the k-neighbour relation $\sim_k$.

Proof. Since formal étaleness is stable under pull-back, we see, using Lemma 17.3, that we may use a formal-étale monic covering of N by models V_j, $j \in J$, to find a formal-étale monic covering of M by models U_i such that f maps each U_i into some V_j. Since $U_i \rightarrowtail M$ creates and $V_j \rightarrowtail N$ preserves $\sim_k$, it is enough to see the Proposition for the restriction of f to U_i, which is a map $\varphi \colon U_i \to V_j$. Using monic formal-étale inclusions $U_i \hookrightarrow R^m$, $V_j \hookrightarrow R^n$ (which create $\sim_k$, by Lemma 17.3), the result follows from Corollary 6.5.

In naive terms, the Proposition can be stated:

$\forall x \in M$: $f(M_k(x)) \subseteq M_k(f(x))$.

Proposition 17.6. Any formal manifold M is infinitesimally linear, and satisfies condition W and the symmetric functions property.

Proof. We prove this first for the model objects $U \subseteq R^n$. The proofs are so similar that we shall only give it for the property W, and we give it naively, only. Let $\tau: D \times D \to U$ be constant on the axes. By property W for R^n, there exists a unique t: $D \to R^n$ making

$$\begin{array}{ccc} D \times D & \xrightarrow{\cdot} & D \\ \tau \downarrow & & \downarrow t \\ U & \rightarrowtail & R^n \end{array}$$

commute; but by formal-étaleness of $U \rightarrowtail R^n$, t factors through U.

For a general formal manifold M, suppose that $\tau: D \times D \to M$ is constant on the axes, with value $x \in M$, say. Let $\varphi: U \rightarrowtail M$ be monic formal-étale with U a model and x in the image of φ. Since $D \times D$ is infinitesimal and φ is formal-étale, τ factors across φ, $\tau = \varphi \circ \tau'$, and using property W for U, we get a factorization of τ' over $D \times D \xrightarrow{\cdot} D$, hence also of τ. The uniqueness follows because any factorization of $\tau: D \times D \longrightarrow M$ over $D \times D \xrightarrow{\cdot} D$ will map 0 into x, and hence by formal-étaleness of φ factor through U; now use the uniqueness assertion of property W for U.

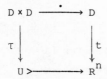

EXERCISES

17.1. Assume $f: M \to N$ is $\mathcal{D}$-étale. Prove that for any $X \in \mathcal{E}$, $f^X: M^X \to N^X$ is $\mathcal{D}$-étale.

17.2. Assume $\mathcal{E}$ has the exactness property that jointly epic families are stable under pull-back (cf. Appendix B) (every topos has this property). Prove that if $1 \to D$ is a map in $\mathcal{D}$, and $\{U_i \to M \mid i \in I\}$ is a jointly epic class of $\mathcal{D}$-étale maps, then the class $\{U_i^D \to M^D \mid I \in I\}$ is jointly epic.

The result is reflected into the following naive argument: "If $\{U_i \hookrightarrow M \mid i \in I\}$ is a covering by formal-étale maps, any tangent vector $t: D \to M$ is a tangent of one of the U_i's; for since the family is covering, $t(0) \in U_i$ for some $i \in I$, and thus since $U_i \to M$ is étale, t factors across it".

17.3. Assume exactness properties like in Exercise 17.2, or equivalently, validity of the mode of reasoning presented in naive terms there. Prove that if U is a model object, then so is TU (and its dimension is the double as that of U).

Prove also that if M is a formal manifold, then so is TM.

17.4. Assume exactness properties like in Exercise 17.2. Let M be a formal manifold of dimension n. Prove that for $x \in M$, the R-module T_xM is isomorphic to R^n, and in particular satisfies the vector form of Axiom 1^W.

In non-naive formulation: $TM \to M$, as an R-module object in $\mathcal{E}/M$ is locally isomorphic to R^n (meaning to $R^n \times M \to M$).

I.18: DIFFERENTIAL FORMS IN TERMS OF 1-NEIGHBOUR SIMPLICES

Whenever we have a formal manifold M, we may use the binary relation $\sim_1$ to form a simplicial object

$$M \leftleftarrows M_{(1)} \begin{smallmatrix}\longleftarrow\\ \longleftarrow\\ \longleftarrow\end{smallmatrix} M_{(1,1)} \begin{smallmatrix}\longleftarrow\\ \longleftarrow\\ \longleftarrow\\ \longleftarrow\end{smallmatrix} \cdots \tag{18.1}$$

where $M_{(1)} \subseteq M \times M$ is as considered in the previous §, $M_{(1,1)} \subseteq M{\times}M{\times}M$ is given by

$$M_{(1,1)} = [\![(x,y,z) \mid x \sim_1 y \wedge y \sim_1 z \wedge z \sim_1 x]\!]$$

etc: ($M_{(1,\ldots,1)} = [\![(x_1,\ldots,x_n) \mid x_i \sim_1 x_j \quad \forall i,j]\!]$). The elements of M, $M_{(1)}$, and $M_{(1,1)}$ may be visualized as 0,1, and 2-simplexes

• , •—• , and  ,

where the lines indicate the relation of being 1-neighbours. The "face" operators ∂_i, i.e. the maps appearing in (18.1), are the operators "omitting the i'th vertex". A <u>degenerate</u> simplex is one in which two vertices are equal, e.g. (x,y,x) is a degenerate 2-simplex, and (x,y,z,y) is a degenerate 3-simplex.

Note that there is an evident diagonal map $\Delta: M \to M_{(1)}$ given by $x \longmapsto (x,x)$ (and in higher dimensions, there are many; they are the "degeneracy operators").

We assume in the following Axiom 1^W.

<u>Theorem 18.1</u>. Given a map $h: M \to N$ between formal manifolds. Then there is a bijective correspondence between

maps $\bar{h}: M_{(1)} \longrightarrow N$ with $\bar{h} \circ \Delta = h$ (i)

fibrewise R-linear maps $H: TM \longrightarrow TN$ over h (ii)

(meaning H maps $T_x M$ linearily to $T_{h(x)} N$ for $\forall x \in M$; both of these tangent spaces are R-modules, since formal manifolds are infinitesimally linear).

<u>Proof</u>. We shall give the correspondence only for the case where M and N are model objects; the rest is straightforward patching. So assume $M \subseteq R^m$, $N \subseteq R^n$ with the inclusions being formal-étale. Then we have isomorphisms

$$M_{(1)} \longrightarrow M \times D(m)$$

$$(x,y) \longrightarrow (x, y-x)$$

and

$$M^D = TM \xrightarrow{<\pi, \gamma>} M \times R^m$$

as well as

$$N^D = TN \xrightarrow{<\pi, \gamma>} N \times R^n$$

where γ denotes principal-part formation. To abbreviate some terminology in the following, we say that a map $\eta: M' \to N'$ from an object $M' \to M$ over M to an object $N' \to N$ over N is <u>lifting</u> if the diagram

$$\begin{array}{ccc} M' & \xrightarrow{\eta} & N' \\ \downarrow & & \downarrow \\ M & \dashrightarrow & N \end{array}$$

can be completed (dotted arrow) in a commutative way. The dotted arrow is said to be one by means of which η is lifting. By the above isomorphisms, we then have the following string of conversions

$$M_{(1)} \longrightarrow N \quad (1)$$

$$\overline{M \times D(m) \longrightarrow N} \quad (2)$$

$$\overline{M \longrightarrow N^{D(m)}} \quad (3)$$

$$\overline{M \longrightarrow N \times \underset{\sim}{\mathrm{hom}}_{R\text{-lin}}(R^m, R^n)} \quad (4)$$

$$\overline{M \times R^m \longrightarrow N \times R^n \text{ which is lifting and fibrewise linear}} \quad (5)$$

$$\overline{TM \longrightarrow TN \text{ which is lifting and fibrewise linear.}} \quad (6)$$

In each of the 6 stages, there is an obvious condition concerning h, e.g. for (1): the composite with $\Delta: M \to M_{(1)}$ is h, and for (6): the map is lifting by means of h. All these data correspond under the conversions. Note that the passage from (3) to (4) is bijective in virtue of Axiom 1^W (actually Axiom 1" suffices). Thus we get a bijective correspondence between data (i) and (ii). The fact that it does not depend on the formal-étale embeddings of M and N into R^m and R^n, respectively (which is what makes patching possible) follows because we can describe the passage from (i) to (ii) explicitely without using the embeddings: Given $\bar{h}: M_{(1)} \longrightarrow N$. We describe the resulting $H: TM \to TN$ (naively) as follows: given $(t: D \to M) \in TM$, with $\pi(t) = x \in M$. Since $t(0) = x$, $x \sim_1 t(d)$ $\forall d \in D$, so $\bar{h}(x,t(d)) \in N$ makes sense $\forall d \in D$. Now

$$(d \longmapsto \bar{h}(x,t(d))) \in TN$$

is a tangent vector at N, and it sends 0 to $\bar{h}(x,x) = h(x)$, so that its base point is $h(x)$.

Corollary 18.2. Assume that V is an R-module which satisfies the vector form of Axiom 1^W, and let M be a formal manifold. Then there is a natural bijective correspondence between maps $\bar{h}: M_{(1)} \longrightarrow V$ with $\bar{h} \circ \Delta \equiv 0$ (i)

and

differential 1-forms on M with values in V. (ii)

Proof. If V is furthermore a formal manifold, this is a straightforward consequence of the Theorem. In the general case, one must adapt the proof rather than the theorem.

There is a related Theorem for higher forms. The n_o appearing is typically the unit element of a group object G.

Theorem 18.3. Let M and N be formal manifolds and $n_o \in N$ an element, $\mathbf{1} \longrightarrow N$. There is a natural bijective correspondence between

maps $\bar{h}: M_{(1,\ldots,1)} \to N$, taking value n_o on all degenerate simplices (i)

and

maps $TM \times_M \cdots \times_M TM \longrightarrow T_{n_o}N$, which are k-linear and alternating. (ii)

(In (i), the index $(1,\ldots,1)$ contains k 1's; in (ii), there are k factors TM.).

Proof. Again, we first assume that we have formal-étale inclusions $M \subseteq R^m$ and $N \subseteq R^n$; and assume $n_o = \underline{0} \in R^n$. Then by Proposition 16.5,

$$M_{(1,\ldots,1)} \cong M \times \tilde{D}(k,n)$$

so that we have the following conversions:

$M_{(1,\dots,1)} \longrightarrow N$ taking all degenerate simplices to $\underline{0}$

* * *

$M \times \widetilde{D}(k,m) \longrightarrow N$ taking $(x,\underline{\tilde{d}})$ to $\underline{0}$ whenever $\underline{\tilde{d}}$ belongs to any of the k copies of $\tilde{D}(k-1,n)$ in $\tilde{D}(k,n)$

* * *

$M \to [\widetilde{D}(k,m),R^n]$ (recall the notation of (16.7))

* * *

$M \longrightarrow \underset{\sim\sim}{\text{hom}}_{k\text{-lin., alternating}}(R^m \times \dots \times R^m, R^n)$

* * *

$M \times R^m \times \dots \times R^m \to R^n$ fibrewise k-linear alternating

* * *

$TM \times_M \dots \times_M TM \to R^n = T_{\underline{0}}N$ fibrewise k-linear alternating

* * *

The proof of the well-definedness of this passage, i.e. independence of the choice of formal-étale embeddings of M and N (and hence the patching argument) is less immediate than for the case $k = 1$ carried out in the proof of Theorem 18.1, and depends on standard multilinear algebra. We omit it.

<u>Corollary 18.4</u>. Let M be a formal manifold, and V an R-module which satisfies the vector form of Axiom 1^W. Then there is a natural 1-1 correspondence between

maps $\bar{h}: M_{(1,\dots,1)} \longrightarrow V$ taking value $\underline{0}$ on degenerate simplices

and

differential k-forms on M with values in V (differential forms taken in the sense of Definition 14.1).

<u>Proof</u>. This is proved the same way as Corollary 18.2 was proved from Theorem 18.1.

An important example of a module which satisfies Axiom 1^W without being a formal manifold is $\text{Vect}(M)$, the object of vector fields on a formal manifold M. This is a consequence of Exercise 17.4.

We shall finish this § with some remarks without proofs (they can be found in [37]) about a very natural and directly geometric differential-form notion that becomes possible in virtue of the Theorems 18.1 and 18.3 (not their respective Corollaries). Let G be a group which is a formal manifold. It is not assumed commutative, so we write multiplicatively; $e \in G$ is the neutral-element. Let M be a formal manifold. A k-<u>cochain on</u> M <u>with values in</u> G is a map

$$\omega: M_{\underbrace{(1,\ldots,1)}_{k \text{ times}}} \longrightarrow G$$

(not to be confused with the <u>cubical</u> cochains considered in §14), and it is called <u>normalized</u> if its value is e on any degenerate simplex (whence its value on <u>any</u> simplex is "infinitesimal", i.e. 1-neighbour to $e \in G$). Normalized 1-cochains occur naturally in geometry: M might for instance be a curve in physical space, and ω might associate to a pair of neighbouring points on it the rotation $\in SO(3)$ which the Frenet frame gets by going from one of the points to the other.

It is not difficult to prove (by working in coordinates) that for a normalized 1-cochain

$$\omega(x,y) \cdot \omega(y,x) = e \qquad \forall x \sim_1 y \tag{18.2}$$

Given a 1-cochain ω, we may define a 2-cochain $d\omega$ by

$$(d\omega)(x,y,z) := \omega(x,y) \cdot \omega(y,z) \cdot \omega(z,x)$$

(which one might think of as the 'curve integral of ω around the edges of the infinitesimal triangle xyz). From (18.2), it is immediate that if ω is a normalized 1-cochain, then $d\omega$ is a normalized 2-cochain.

Also, if $j: M \to G$ is any function (= 0-cochain), we may define a normalized 1-cochain dj on M by putting

$$(dj)(x,y) := j(x)^{-1} \cdot j(y),$$

and an immediate calculation shows that $d(dj) = 0$, the 2-cochain with constant value $e \in G$.

Furthermore, if $f: N \to M$ is a map between formal manifolds, then since f preserves the relation $\sim_1$, we get immediately maps f^* from the set of k-cochains on M to the set of k-cochains on N, and

$$d(f^*j) = f^*(dj),$$

$$d(f^*\omega) = f^*(d\omega),$$

whenever j, respectively ω, is a 0-, respectively 1-cochain on M.

In particular, consider $i: G \to G$, the identity map on G; it is a G-valued 0-cochain on G. Thus, di is a (normalized) 1-cochain on G, and, for any $j: M \to G$

$$dj = d(i \circ j) = d(j^*i) = j^*(di) \tag{18.3}$$

so that the 1-cochain di on G plays a special universal role.

Now, by Theorem 18.3, for $k = 1$ and 2, normalized 1- and 2-cochains on M with values in G correspond bijectively to differential 1- and 2-forms on M with values in T_eG. Let us by $\overline{\omega}$ denote the differential form corresponding to the normalized cochain ω, and also let us denote the coboundary operator for differential forms by $\overline{d}$. It is then possible (by working in coordinates) to prove the following Comparison Theorem between d and $\overline{d}$. It is not surprising that it should involve the Lie bracket (Exercise 9.3) on T_eG, since the definition of d utilizes the group structure on G, whereas T_eG only remembers that G had a group structure via the Lie bracket.

Theorem 18.5. For any normalized 1-cochain ω,

$$\overline{d\omega} = \tfrac{1}{2}(\bar{d}\,\bar{\omega} + \tfrac{1}{2}[\bar{\omega},\bar{\omega}]).$$

(here, $[\bar{\omega},\bar{\omega}]$ is a certain 2-form on M with values in T_eG defined using the Lie bracket on T_eG; generally

$$[\bar{\omega},\bar{\theta}](u,v) = [\bar{\omega}(u),\bar{\theta}(v)] - [\bar{\omega}(v),\bar{\theta}(u)] ,$$

where $\bar{\omega}$ and $\bar{\theta}$ are 1-forms with values in a Lie algebra, and u and v are tangent vectors at the same point of M).

The universal role of the 1-cochain di on G is transferred to a similar universal role for the 1-form $\overline{di}$ on G with value in the Lie algebra T_eG; $\overline{di}$ is the socalled Maurer-Cartan form Ω. Since $ddi = 0$, the theorem has as immediate Corollary (take $\omega = di$) the Maurer-Cartan formula

$$\bar{d}\,\Omega = -\tfrac{1}{2}[\Omega,\Omega].$$

Several conditions get a more natural statement when expressed in terms of G-valued normalized cochains rather than in terms of Lie-algebra valued forms. For instance, the condition for a T_eG-valued 1-form $\bar{\theta}$ on M to be of form $f^*\Omega$ for some $f: M \to G$ then simply becomes the statement that the corresponding 1-cochain θ is a coboundary $\theta = df$, for some f. For,

$$\theta = df$$

iff

$$\theta = f^*di \qquad \text{(by (18.3))}$$

iff

$$\bar{\theta} = \overline{f^*di} = f^*\overline{di} = f^*\Omega .$$

Some of the notions above, and the comparison theorem, also holds for cochains with values in groups G that are not themselves formal manifolds, but are transformation groups on formal manifolds N. For the full transformation group Diff(N) of all bijective maps $N \to N$, $T_e G$ is the Lie algebra Vect(N) of vector fields on N.

EXERCISES

18.1. Prove that if $G = (R,+)$, then, identifying $T_o R$ with R, the form Ω is just the γ of (1.3).

18.2. Prove that for any group object $(G,\cdot)$ which is a formal manifold, the Maurer-Cartan form Ω may be described by

$$\Omega(t)(d) = t(0)^{-1} \cdot t(d) .$$

I.19: OPEN COVERS

In this §, we introduce a notion of <u>open</u> inclusion in any sufficiently exact category $\mathcal{E}$ equipped with a k-algebra object R; the subobject $\mathrm{Inv}(R) \rightarrowtail R$ "of invertible elements" will be open, and, assuming Axiom 1_k^W for R, also formal-étale. We shall furthermore introduce an Axiom (Axiom 3) which will allow us to conclude: any open inclusion is formal-étale. As a Corollary, we may conclude that certain specific objects constructed out of R, like the projective plane over R, more generally, any Grassmannian, are formal manifolds.

Assume in the rest of this § that $\mathcal{E}$ is a cartesian closed category with finite inverse limits. The subobject $\mathrm{Inv}(R) \rightarrowtail R$ can be defined using the latter; it is the composite

$$[\![(x,y) \in R^2 \mid x\cdot y = 1]\!] \hookrightarrow R \times R \xrightarrow{\ proj_1\ } R,$$

which is always monic.

<u>Proposition 19.1</u>. Assume R satisfies Axiom 1_k^W. Then the subobject $\mathrm{Inv}(R) \rightarrowtail R$ is formal-étale.

<u>Proof</u>. Let $W = (k^n, \mu)$ be a Weil-algebra, and $J = \mathrm{Spec}_R(W)$ the infinitesimal object it defines. We must prove the following square to be a pull-back

$$\begin{array}{ccc} (\mathrm{Inv}(R))^J & \rightarrowtail & R^J \\ \downarrow & & \downarrow \\ \mathrm{Inv}(R) & \rightarrowtail & R \end{array} .$$

Since $(-)^J$ commutes with inverse limits, it commutes with the formation of Inv, so that the question is whether

$$\begin{array}{ccc} \mathrm{Inv}(R^J) & \rightarrowtail & R^J \\ \downarrow & & \downarrow \\ \mathrm{Inv}(R) & \rightarrowtail & R \end{array}$$

is a pull-back. By Axiom 1^W, this diagram in turn is isomorphic to

$$\begin{array}{ccc} \mathrm{Inv}(R \otimes W) & \rightarrowtail & R \otimes W = R^n \\ \downarrow & & \downarrow \mathrm{proj}_1 \\ \mathrm{Inv}(R) & \longrightarrow & R \end{array}$$

To see that this is a pull-back amounts to proving that if $\underline{a} = (a_1,\ldots,a_n) \in R \otimes W$ has a_1 invertible in R, then $\underline{a}$ is invertible in $R \otimes W$. We may divide through by a_1, or equivalently, let us assume $a_1 = 1$. Let $\underline{u}$ denote $-(0,a_2,\ldots,a_n)$. Then since W is a Weil algebra, $\underline{u}^n = 0$. Then $\underline{a} = 1-\underline{u}$, and

$$(1-\underline{u})\cdot(1 + \underline{u} + \underline{u}^2 + \ldots + \underline{u}^{n-1}) = 1$$

so $\underline{a}$ is invertible.

We shall consider some stability properties which the class of formal-étale maps has (under suitable assumptions on E); in conjunction with formal-étaleness of $\mathrm{Inv}(R) \rightarrowtail R$, this will give us many new formal-étale maps.

Let U denote the class of formal-étale maps. Then:

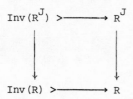

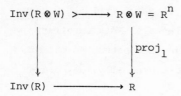

Proposition 19.2.

(i) U is closed under composition and contains all isomorphisms.

(ii) U is stable under pull-backs (cf. (17.2)).

Proof. Both follow from well known diagrammatic lemmas about pull-backs, and the fact that any functor $(-)^J$ (for any object J) preserves pull-backs.

For the stability properties of U which involve colimits, we need more exactness properties of $\mathcal{E}$ (satisfied whenever $\mathcal{E}$ is a topos), and

Axiom 3_k. For any Weil algebra W over k, the functor $(-)^J$: $\mathcal{E} \longrightarrow \mathcal{E}$ has a right adjoint (where $J = \mathrm{Spec}_R W$).

We express this property of J by saying: J is an atom. Note that $(-)^J$ always has a left adjoint, $J\times -$. The existence of a right adjoint is 'amazing': In the category of sets, only $J = 1$ has this property. Note also that the axiom implies that $(-)^J$ preserves epics and colimits.

The category of Sets (in fact any topos) has the following exactness property: given a commutative diagram

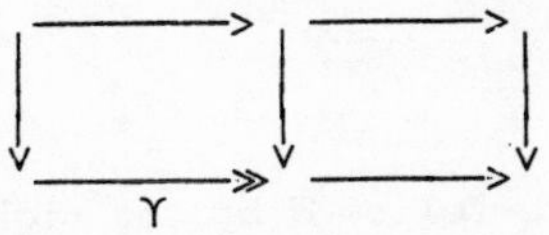

in which the left hand square and the composite square are pull-backs, and γ is epic, then the right hand square is a pull-back.

Proposition 19.3. Assume this exactness property for $\mathcal{E}$, and assume Axiom 3. Then we have the following stability property for the class U of formal-étale maps:

(iii) Given a pull-back square

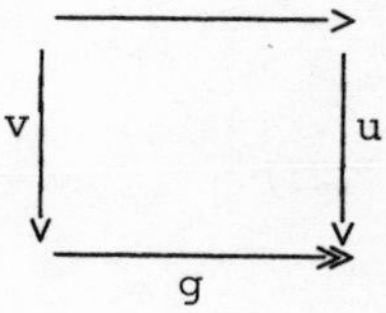

with g epic and $v \in U$; then $u \in U$ ("U descends").

<u>Proof</u>. As in the previous proof, let $J = \mathrm{Spec}_R W$, and consider the box

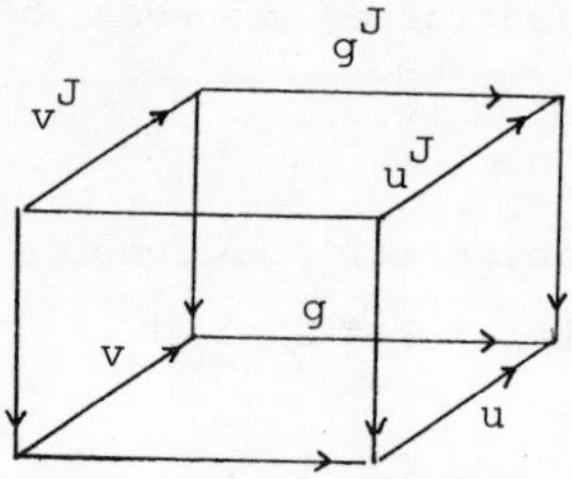

Since $(-)^J$ preserves epics by Axiom 3, g^J is epic. The left hand square is a pull-back since v is formal-étale, and the bottom is a pull-back by assumption. Hence the total diagram

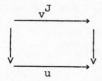

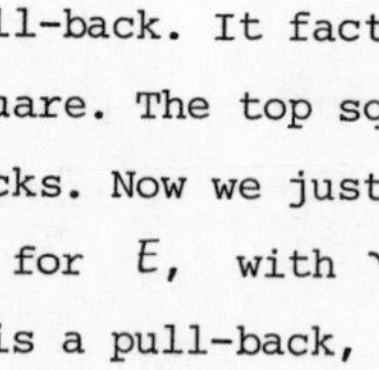

is a pull-back. It factors as the top square followed by the right hand square. The top square is a pull-back since $(-)^J$ preserves pull-backs. Now we just have to invoke the exactness property assumed for E, with $\Upsilon = g^J$, to conclude that the right hand square is a pull-back, which is the desired formal-étaleness property for u with respect to J.

Assuming further "set-like" exactness properties of E (in particular, if E is a topos), it is possible, in the same spirit as in the proof of Proposition 19.3, to prove that Axiom 3 implies that the class U of formal-étale maps, besides the stability properties (i), (ii) and (iii) of Propositions 19.2 and 19.3 also has the properties:

(iv) The epi-mono factorization of a map in U has each of the two factors in U.

(v) If $g \circ p \in U$, $p \in U$, and p is epic, then $g \in U$.

(vi) The inclusions into a coproduct

$$A_i \overset{incl_i}{\rightarrowtail} \Sigma A_i$$

belongs to U; and a map $f\colon \Sigma A_i \to B$ belongs to U if each $f \circ incl_i$ does.

(vii) If $f\colon A \to B \in U$, then $\Delta\colon A \longrightarrow A \times_B A \in U$.

The class of stability properties (i)-(vii) were considered by Joyal to define the concept of an <u>abstract étaleness notion.</u> The class of formal-étale maps thus is an abstract étaleness notion, and it contains $Inv(R) \rightarrowtail R$. The smallest abstract étaleness notion containing this map may be called the class of <u>(strongly)</u> <u>étale</u> maps; the monic strongly étale maps are called <u>open inclusions</u>. What is important is that the natural atlases in algebraic geometry (for Grassmannians relative to R, say) are open inclusions with domain R^k for suitable k. These open coverings of the Grassmannians, being coverings by formal-étale maps from R^k, allow one to conclude that the Grassmannians are formal manifolds. (See e.g. [42] ; it contains a weaker theorem, but the proof will easily carry the stronger result also.)

EXERCISES

19.1. Prove that the class U of formal-étale maps has the properties (iv)-(vii) (for $\mathcal{E}$ a topos). Note: The proof of (v) may be found in [36]; Lemma 3.3, the second assertion in (vi) may be found in [42], Lemma 4.6.

19.2. Let $R/\equiv$ denote the set of orbits of the multiplicative action of $\mathrm{Inv}(R)$ on R. Assume Axiom 1^W and Axiom 3. Prove that "$R/\equiv$ believes that the addition map $D \times D \to D_2$ is surjective", i.e. that if $\bar{f}_i : D_2 \to R/\equiv$ $(i = 1,2)$ has

$$\bar{f}_1(d_1+d_2) = \bar{f}_2(d_1+d_2) \quad \forall\ (d_1,d_2) \in D \times D, \tag{19.1}$$

then $\bar{f}_1 = \bar{f}_2$.

Hint (for a naive argument). Use that D_2 is an atom to lift the $\bar{f}_i$ $(i = 1,2)$ to maps $f_i : D_2 \to R$. Use next the assumption (19.1) and the fact that $D \times D$ is an atom to find $h: D \times D \to \mathrm{Inv}(R)$ with

$$f_1(d_1+d_2) = h(d_1,d_2) \cdot f_2(d_1+d_2) \quad \forall\ (d_1,d_2) \in D \times D.$$

Prove that h may be chosen symmetric and with $h(0,0) \in \mathrm{Inv}(R)$. Use symmetric functions property for R and formal étaleness for $\mathrm{Inv}(R)$ to find $k: D_2 \to \mathrm{Inv}(R)$ with $h(d_1,d_2) = k(d_1+d_2)$ $\forall (d_1,d_2) \in D \times D$. Use symmetric functions property for R to conclude

$$f_1(\delta) = k(\delta) \cdot f_2(\delta) \qquad \forall\, \delta \in D_2,$$

and conclude $\bar{f}_1 = \bar{f}_2$ on D_2.

19.3. Prove that the product of two atoms is an atom.

19.4. Prove that if D is an atom and B and X are arbitrary objects, then $(X_D)^B \cong (X^{(B^D)})_D$. (Hint: the two functors $\mathcal{E} \rightarrow \mathcal{E}$ given by

$$(-)^D \circ (- \times B) \quad \text{and} \quad (- \times B^D) \circ (-)^D$$

are isomorphic; now take their right adjoints.)

I.20: DIFFERENTIAL FORMS AS QUANTITIES

The Axiom 3 introduced in the previous § is very radical: whereas Axiom 1 in its various forms, and its generalization, Axiom 2, lead to a reformulation of the classical differential concepts (which is close to the reasoning employed by Lie and others), Axiom 3 leads into new "previously undreamed-of" land where a "drastic simplification of the usual differential form calculus may be in the offing" (to quote [51]).

The simplification arises from the fact that Axiom 3 allows one to replace differential n-forms on M, which are functionals, i.e. functions defined on function spaces (namely the M^{D^n}), by certain functions or "quantities" defined on M itself,

$$M \longrightarrow \Lambda^n,$$

where the codomain is a highly non-classical object, constructed in virtue of Axiom 3. (Since it classifies the cochains of the deRham complex, it plays a role analogous to Eilenberg-MacLane complexes $L(\pi,n)$ in simplicial algebraic topology.)

Let E be a cartesian closed category. We say that an object $J \in E$ is an atom if the functor $(-)^J\colon E \to E$ has a right adjoint, which we denote $(-)_J$. Besides the usual "λ-conversion" rule

$$\frac{A \times J \longrightarrow B}{A \longrightarrow B^J} \tag{20.1}$$

we thus have the further conversion

$$\frac{B^J \longrightarrow C}{B \longrightarrow C_J} . \qquad (20.2)$$

Whereas the categorical logic, as we presently know it, and as it is exposed in Part II, is well suited to express the λ-conversion (20.1) in set-theoretic terms, it fails completely for the rule (20.2). An aspect of the reason for this failure is that in the category of sets, $J = \mathbf{1}$ is the only atom. (More generally, in toposes, (20.1) is an indexed, or <u>locally internal</u> adjointness, [28] (Appendix), whereas (20.2) is not, unless $J = \mathbf{1}$.)

Let now $\mathcal{E}$ be a cartesian closed category with finite inverse limits, and let R be a k-algebra object in $\mathcal{E}$ satisfying Axiom 3.

Let V be an object on which $(R,\cdot)$ acts (in the applications, V is an R-module, typically R itself). By (20.2), there is a 1-1 correspondence between (singular infinitesimal cubical) n-cochains on an object M, with values in V,

$$M^{D^n} \xrightarrow{\ \omega\ } V$$

and maps

$$M \xrightarrow{\ \hat{\omega}\ } V_{D^n} .$$

<u>Proposition 20.1</u>. There exists a subobject $\Lambda^n(V) \subseteq V_{D^n}$ such that for any M, ω , as above, $\hat{\omega}$ factors through $\Lambda^n V$ if and only if ω is a differential n-form (in the sense of Definition 14.2).

<u>Proof</u>. We give it for the case $n = 1$ only. We construct two maps $a,b: V_D \longrightarrow (V_D)^R$, and take $\Lambda^1(V)$ to be their equalizer; we construct them by constructing for each $X \in \mathcal{E}$, two maps (natural in X)

$$\hom_{\mathcal{E}}(X,V_D) \longrightarrow \hom_{\mathcal{E}}(X,(V_D)^R),$$

and apply Yoneda's lemma. Now, the domain object here is identified via (20.2) with

$$\hom_{\mathcal{E}}(X^D, V),$$

and the codomain object (via (20.1) and (20.2)) with

$$\hom_{\mathcal{E}}((X \times R)^D, V),$$

so we should construct processes $\bar{a}$ and $\bar{b}$ leading from V-valued 1-cochains on X to V-valued 1-cochains on $X \times R$. The process $\bar{a}$ assigns to a 1-cochain $\theta\colon X^D \to V$ the 1-cochain on $X \times R$ which to

$$D \xrightarrow{\langle t, \lambda \rangle} X \times R \tag{20.3}$$

associates

$$\theta(\lambda(0) \cdot t) \in V.$$

The process $\bar{b}$ assigns to θ the 1-cochain on $X \times R$ which to (20.3) associates

$$\lambda(0) \cdot \theta(t) \in V.$$

(Both these descriptions of $\bar{a}$ and $\bar{b}$ are naive, but they fall under the scope of the categorical logic of Part II, since no right adjoints $(-)_D$ are involved for $\bar{a}$ and $\bar{b}$). Now it is easy to see that $\omega\colon M^D \to V$ is a 1-form if and only if $\bar{a}(\omega) = \bar{b}(\omega)\colon (M \times R)^D \to V$. For, if $\bar{a}(\omega) = \bar{b}(\omega)$, we get in particular two equal composites

$$M^D \times R \xrightarrow{1 \times \Delta} M^D \times R^D \cong (M \times R)^D \underset{\bar{b}(\omega)}{\overset{\bar{a}(\omega)}{\rightrightarrows}} V$$

which is the desired homogeneity condition. Conversely, the homogeneity condition implies the equality $\bar{a}(\omega) = \bar{b}(\omega)$, because both these as maps $M^D \times R^D \to V$ factor across

$$M^D \times R^D \xrightarrow{1\times\pi} M^D \times R$$

(π = evaluation at $0 \in D$), by construction.

Thus $\omega: M^D \to V$ is a 1-form iff $\bar{a}(\omega) = \bar{b}(\omega)$, which is the case iff $a(\hat{\omega}) = b(\hat{\omega})$ (by construction of a and b in terms of $\bar{a}$ and $\bar{b}$), that is, iff $\hat{\omega}$ factors across the equalizer of a and b.

For the case where $V = R$, we write Λ^n for $\Lambda^n(V)$. It represents, by the Proposition, differential n-forms,

$$\hom_E(M,\Lambda^n) \cong \underline{\text{set}} \text{ of differential n-forms on } M, \qquad (20.4)$$

but we do not have $(\Lambda^n)^M \cong$ object of differential n-forms; see Exercise 20.1.

The exterior derivative for differential n-forms introduced in Theorem 14.5 is natural in M, and thus, by Yoneda's lemma and (20.4), is represented by a map

$$\Lambda^{n-1} \xrightarrow{d} \Lambda^n .$$

We shall describe this d explicitely for $n = 1$. Clearly, $\Lambda^o = R$. The d in question is the map

$$R \xrightarrow{d} \Lambda^1 \subseteq R_D$$

corresponding under (20.2) to the "principal-part" map (1.3) or (7.4):

$$R^D \xrightarrow{\gamma} R.$$

For, if we denote the conversion (downwards) in (20.2) by $^\wedge$, we have, for any $f: M \to R$

$$(df)^\wedge = (\gamma \circ f^D)^\wedge = \hat{\gamma} \circ f ,$$

the first equality sign by (14.14), and the second one by naturality of (20.2) in the variable B.

Thus, if we rethink the differential-form notion via the Λ^n's, we have in particular

$$df = d \circ f$$

(where $d = \hat{\gamma}$), which surely is an amazing way to get the differential (or gradient) df of a function f. Note also that for ω an n-form on M, and $g: N \to M$,

$$g^*\omega = \omega \circ g,$$

as soon as we consider forms as maps into Λ^n ("quantities with values in Λ^n").

EXERCISES

20.1. For $n \geq 1$, the only n-form on **1** is the form 0. So the <u>object</u> of n-forms on **1** is **1**. Conclude that $(\Lambda^n)^{\mathbf{1}}$ ($\cong \Lambda^n$) is not the object of n-forms on **1**.

I.21: PURE GEOMETRY

It is natural to ask which axiomatizations of, say, projective geometry are compatible with the present theory ? Let us first look at plane projective geometry. Besides the primitive notions 'point' and 'line', we would put a binary relation 'distinct' as a primitive. It is incompatible with our theory to have "non-distinct implies equal" even though it is compatible to have "non-equal implies distinct". In particular, "equal" will be stronger than "not unequal".

Basic axioms are then "two distinct points determine a unique line", and "two distinct lines intersect in a unique point". In Hjelmslev's work (as quoted in §1), two points always are connected by <u>at least</u> one line (which is unique iff the points are distinct), whereas, in our context, we do not want to assert even the existence of a connecting line, except for a pair of <u>distinct</u> points. We can argue for this feature in the affine plane $R \times R$ as follows. Let $(d_1,d_2) \in D \times D$. If we had a line through $(0,0)$ and (d_1,d_2), say the line with slope α relative to the x-axis, then $d_2 = \alpha \cdot d_1$, and hence $d_1 \cdot d_2 = 0$; similarly if $(0,0)$ and (d_1,d_2) lie on a line with a slope relative to the y-axis. In both cases $d_1 \cdot d_2 = 0$, so we arrive at the conclusion $D(2) = D \times D$, which contradicts Axiom 1, cf. Exercise 4.6.

So, also in the <u>projective</u> plane we give up "any two points are connected by at least one line" as well as the dual "any two lines have at least one point in common".

If we now turn to projective <u>space</u>, we have of course the phenomenon of mutually skew lines ℓ and m (every point on ℓ is distinct from every point on m), so they do not intersect, but for a more positive reason than that which may cause two lines in the projective plane not to have any point in common. It becomes imperative to have a notion "two lines intersect" which is weaker than "having a point in common". So "ℓ and m intersect" is not going to mean

$$\exists P: P \in \ell \wedge P \in m, \tag{21.1}$$

but rather

"the volume of the tetrahedron spanned by any two distinct points on ℓ and any two distinct points on m is zero" (21.2)

Now, of course, 'volume' of tetrahedra is not well-defined in the projective space, but it is well defined modulo multiplication by invertible scalars.

More precisely, let $R/\equiv$ denote R modulo the action of the multiplicative group $\mathrm{Inv}(R)$ of units in R. It contains at least two distinct elements $\{0\}$ and $\{1\}$, but it also contains things like $\{d\}$, for any $d \in D$ (equivalence classes, or orbits, being denoted by curly brackets). For any two lines ℓ and m in projective space, the volume of the tetrahedron spanned by two distinct points on ℓ and two distinct points on m is well-defined as an element of $R/\equiv$; let us denote it $S(\ell,m) \in R/\equiv$. So

$$\ell \text{ and } m \text{ interesect} \underset{\text{Def.}}{\Longleftrightarrow} S(\ell,m) = \{0\} . \tag{21.2'}$$

With this definition, any two coplanar lines intersect. It is clear that (21.1) implies (21.2).

Having $R/\equiv$ (a kind of "truth value object" for geometry), we try to see to what extent some synthetic modes of speaking in classical differential geometry become captured: - in particular, we ask whether it is true that "the neighbouring generators of a ruled surface are in general mutually skew, but there exist a special kind of ruled surface in which they "intersect", namely the developpables [Torsen]"(Klein [29], p.113).? Klein explains what he means by "intersect" in analytic terms: "When two generators of a developpable ruled surface converge, then their shortest distance becomes infinitely small in higher than first order". With the intersection notion we introduced above, this is paraphrased: On any 1-parametrized family $\ell(t)$ of lines in projective 3-space, $S(\ell(t),\ell(t+d)) = \{0\}$ $\forall d \in D$, but only for the developpables do we have $S(\ell(t),\ell(t+d)) = \{0\}$ $\forall d \in D_2$. Calculations with Plücker coordinates substantiate the correctness of the paraphrasing.

When axiomatizing projective space geometry, we would, besides the distinctness relation, and the incidence relation (for points on lines and planes) introduce the 'interesection' relation (21.2) for lines in space as a primitive. However, for many arguments of geometrical-combinatorial nature, the stronger 'incidence-theoretic' condition (21.1) can be used instead, as will be illustrated now.

It is a geometric fact that the family of tangents of a space curve form a developpable. How do we prove this synthetically? Let the space curve be parametrized by the line R,

$$R \xrightarrow{K} \mathbb{P}^3 .$$

We are required to prove

$$S(\ell(t),\ell(t+\delta)) = \{0\} \quad \forall \delta \in D_2, \tag{21.3}$$

where $\ell(s)$ denotes the line in $\mathbb{P}^3$ which is tangent to the curve K at the point $K(s)$. Now $R/\equiv$ has the property that it believes the addition map $D \times D \to D_2$ is surjective (cf. Exercise 19.2), i.e., it suffices to prove (21.3) for all δ of form d_1+d_2 with $(d_1,d_2) \in D \times D$. Now consider the point $K(t+d_1)$. Since $K(t) \sim_1 K(t+d_1)$, we have

$$K(t+d_1) \in \ell(t)$$

(the tangent to K in $K(t)$ may be defined as the unique line containing $K(t)$ and all its 1-neighbours on the curve). Similarly, since $K(t+d_1) \sim_1 K(t+d_1+d_2)$, we have

$$K(t+d_1) \in \ell(t+d_1+d_2).$$

But this means that the two lines $\ell(t)$ and $\ell(t+\delta)$ (with $\delta = d_1+d_2$) have a point, namely $K(t+d_1)$, in common; so (21.1) and hence (21.2) holds.

Leaving the problem of lifting $R/\equiv$-valued functions over R aside, one can similarly make use of the principle, satisfied in many models: "R believes $\mathrm{Inv}(R) \hookrightarrow R$ is surjective", or the weaker "R believes $\mathrm{Inv}(R) \cup \{0\} \hookrightarrow R$ is surjective". In this way one recaptures the way of reasoning often applied in "statements with truth value in $R/\equiv$": Namely that, for a statement involving two points, one may assume these distinct (respectively, may assume them distinct <u>or</u> equal). All incidence statements in geometry are such statements with "quantitative truth values".

PART II

CATEGORICAL LOGIC

INTRODUCTION

It was stressed in the introduction to Part I that the Axioms and concepts of synthetic differential geometry need to be interpreted in other categories than the category of sets (notably in certain Grothendieck toposes).

To this end, and because it in itself crystallizes some ideas, we formulate them in diagrammatic terms, like when we express Axiom 1 by saying that a certain map $R \times R \to R^D$ is invertible. Such formulations amount from a syntactic viewpoint to formulating things without use of variable-symbols.

Thus, for instance, the associativity of the addition $+$ on R is expressed variable-free by saying that

$$\begin{array}{ccc} R \times R \times R & \xrightarrow{R \times +} & R \times R \\ {\scriptstyle + \times R}\downarrow & & \downarrow{\scriptstyle +} \\ R \times R & \xrightarrow[+]{} & R \end{array}$$

is commutative, and expressing the other algebraic conditions on R similarly, we arrive at the well-known notion of ring-object in a category with finite products.

However, the use of variables is so useful a tool that one does not want to abandon it even when working in categories. Variables range over sets of elements. What are the elements of the objects in a category $\mathcal{E}$, and how are statements involving variables interpreted systematically ?

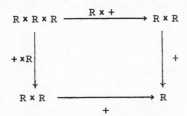

The answer is by no means new, but it is not easy to find a reasonable account, so we try to give one. It will be essential for Part III.

Definition 1.1. Let R be an object in a category E. An element of R is an arrow (map) in E with codomain R. The domain of the map is called the stage of definition of the element.

If r is an element of R, defined at stage X

$$X \xrightarrow{r} R,$$

we also write $r \in_X R$, or $\vdash_X r \in R$. One may even write $r \in R$ if there is no confusion.

If $\alpha: Y \to X$ and $r \in_X R$, we get an element

$$r \circ \alpha \in_Y R,$$

also denoted $\alpha^*(r)$, which we think of as "r, but considered at the 'later' stage Y". An important abuse of notation is to write r instead of $\alpha^*(r)$, thus omitting the change-of-stage map α from notation.

If the category E has a terminal object **1**, then an element r of R defined at stage **1** is called a global element of R (or a global section of R). Given any $Y \in E$, we have the element $r \circ \alpha \in_Y R$, (where $\alpha: Y \to \mathbf{1}$ is the unique such map); or, by the abuse of notation just introduced, $r \in_Y R$. (This is the reason for the name global element: we have the element r at all stages Y.)

If R is a ring object (say) in a category E with finite products, then it is well known (see e.g. [55] III.6, replacing the word 'group' with the word 'ring') that the ring object structure on R gives rise, for each $X \in \mathcal{E}$, to a ring structure on the set $\hom_{\mathcal{E}}(X,R)$. The additive neutral element of R is a global element

$$1 \xrightarrow{0} R$$

For each $X \in \mathcal{E}$, it gives rise to $0 \in_X R$ (omitting the change of stage $X \to 1$ from notation), and this is the additive neutral element in the ring $\hom_{\mathcal{E}}(X,R)$. (In fact, for any $\alpha: Y \to X$, $\alpha^*: \hom_{\mathcal{E}}(X,R) \to \hom_{\mathcal{E}}(Y,R)$ is a ring homomorphism.) Thus, the abuse of notation 'omitting the change of stage' is related to the standard convention of denoting by the same symbol 0 the zero elements of all rings.

We note that if $\mathcal{E}$ = <u>Set</u>, and $R \in \mathcal{E}$, then there is a bijective correspondence between the set R and the set of global elements of the object R: to $r \in R$, associate that unique map

$$1 \xrightarrow{\ulcorner r \urcorner} R$$

which sends the unique element of 1 (= the one-element set) to r. Thus, the element-notion of Definition 1.1 is more general than the ordinary one. Therefore, one sometimes calls elements in the sense of Definition 1.1 for <u>generalized</u> elements. (Perhaps, '<u>parametrized</u> elements' would be better; the stage-of-definition, X, is then to be viewed as the domain of parameters.)

There exists a technique by which one can assume that any given (generalized) element $r: X \to R$ is a global element, namely by replacing the category $\mathcal{E}$ by the comma-category $\mathcal{E}/X$ of objects-over-X; this works provided the 'pull-back functor' $\mathcal{E} \longrightarrow \mathcal{E}/X$ preserves the structure in question. We return to this method in §6.

Note that there is a natural bijective correspondence between

elements of $A \times B$ (1.2)

and

pairs of elements of A and B, with common stage of definition (1.3)

Namely, to $c: X \to A \times B$, associate the pair of elements $\text{proj}_1 \circ c$, $\text{proj}_2 \circ c$. Similarly for several factors.

II.2: SATISFACTION (1)

We describe here how one talks about (generalized) elements by means of mathematical language. To be specific, let us assume that R carries some algebraic structure, say that of a commutative ring object.

If $a,b,c,\ldots$ now are elements of R, defined at same stage, X, say, then we know what we mean by saying

$$a^2 \cdot b + 2c = 0,$$

say, (and similarly for other polynomial equations with integral coefficients), simply because a,b, and c are elements in the ordinary ring $\hom_E(X,R)$. We shall write

$$\vdash_X a^2 \cdot b + 2c = 0 \tag{2.1}$$

to remind ourselves that the elements a,b, and c are defined at stage X, and to make it conform with similar notation to be introduced now for more complicated formulas.

We read '$\vdash_X$' as 'at stage X, the following is <u>satisfied</u>'.

The satisfaction relation $\vdash$ is now defined by induction: suppose $X \in E$ and suppose $\varphi(x)$ is a mathematical formula, for which we, for any $\alpha\colon Y \to X$ and any element $b \in_Y R$ have already defined what we mean by

$$\vdash_Y \varphi(b) \tag{2.2}$$

Then we define

$$\vdash_X \forall x\ \varphi(x)$$

to mean that for any $\alpha: Y \to X$ and any $b \in_Y R$, (2.2) holds. Note that α will occur implictely in (2.2), even though it may not be visible due to the abuse of notation consisting in omitting change of stage from notation.

We give an example. Assume for the sake of the argument that R is a non-commutative ring object. We have, for any $a: X \to R$ (i.e. $a \in_X R$) the formula

$$\varphi(x) = \text{"}x \text{ commutes with } a\text{"} \tag{2.3}$$

For $\alpha: Y \to X$ and $b \in_Y R$, we know what we mean by saying $\varphi(b)$, namely "b commutes with a (more precisely, with $a \circ \alpha$) in the ring $\hom_E(Y,R)$". Now consider the formula

$$\vdash_X \forall x\ \varphi(x),$$

that is, the formula

$$\vdash_X \forall x: \ x \text{ commutes with } a.$$

By the above definition of $\vdash \forall x$, this means

$$\forall \alpha : Y \to X \quad \forall b: Y \to R, \quad a \cdot b = b \cdot a \quad \text{in } \hom_E(Y,R).$$

(note again: a here really stands for $a \circ \alpha$). So to say

$$\vdash_X a \text{ is central}$$

means more than saying that a is a central element in the ring $\hom_E(X,R)$; it means that it <u>remains</u> central in all $\hom_E(Y,R)$, i.e. at all later stages $\alpha: Y \to X$.

As a particular case of the satisfaction relation for the universal quantifier defined above, we note the following. If we can make sense to $\vdash \varphi(b)$ for no matter which stage b is defined at, then the meaning of

$$\vdash_1 \forall x \varphi(x)$$

is simply that $\vdash_Y \varphi(b)$ is true for all Y and all elements b defined at stage Y, or simply: $\vdash \varphi(b)$ holds for all generalized elements, no matter what stage.

We next consider the logical construction $\exists!$ ("there exists a unique"). Suppose $X \in E$, and suppose $\varphi(x)$ is a mathematical formula for which we, for any $\alpha: Y \to X$ and any element $b \in_Y R$ have already defined what we mean by

$$\vdash_Y \varphi(b) \tag{2.4}$$

Then we write

$$\vdash_X \exists!\, x\, \varphi(x)$$

if for any $\alpha: Y \to X$, there exists a unique $b \in_Y R$ for which (2.4) holds.

Finally, suppose $X \in E$, and suppose φ and ψ are mathematical formulae for which we, for any $\alpha: Y \to X$, have already defined what we mean by

$$\vdash_Y \varphi \quad \text{as well as by} \quad \vdash_Y \psi \, .$$

Then we write

$$\vdash_X (\varphi \Rightarrow \psi)$$

if for any $\alpha: Y \to X$ so that $\vdash_Y \varphi$, we also have $\vdash_Y \psi$; and we write

$$\vdash_X (\varphi \wedge \psi)$$

if $\vdash_X \varphi$ and $\vdash_X \psi$. Also, of course,

$$\vdash_X (\varphi \Leftrightarrow \psi)$$

is defined to mean

$$\vdash_X (\varphi \Rightarrow \psi) \wedge (\psi \Rightarrow \varphi).$$

The relation (2.1) has the property: for any $\alpha: Y \to X$,

$$\vdash_X a^2 \cdot b + 2c = 0 \quad \text{implies} \quad \vdash_Y (\alpha^*(a))^2 \cdot \alpha^*(b) + 2\alpha^*(c) = 0$$

because $\alpha^*: \hom_{\mathcal{E}}(X,R) \to \hom_{\mathcal{E}}(Y,R)$ is a ring homomorphism. In the abuse of notation where we omit the α^*, this reads more simply: for any $\alpha: Y \to X$

$$\vdash_X a^2 \cdot b + 2c = 0 \quad \text{implies} \quad \vdash_Y a^2 \cdot b + 2c = 0.$$

We express this property by saying that the formula $a^2 \cdot b + 2c = 0$ is <u>stable</u>. Generally, a formula φ is called <u>stable</u> if $\vdash_X \varphi$ and $\alpha: Y \to X$ imply $\vdash_Y \varphi$. The following is now a trivial observation

<u>Proposition 2.1</u>. For any formulas φ and ψ, the formulas

$$\forall x\, \varphi(x), \qquad \exists!\, x\, \varphi(x), \qquad \varphi \Rightarrow \psi$$

are stable; and if φ and ψ are stable, then so is $\varphi \wedge \psi$.

If we are dealing with more than one object, say a ring object R and a module object V over it, we may indicate what objects the elements and variables are intended to range over in the standard way, e.g. by writing

$$\vdash_1 \forall a \in R \quad \forall u \in V \quad \forall v \in V: a \cdot (u+v) = a \cdot u + a \cdot v$$

for one of the distributive laws.

What would have happened if we in the last formula had written $\forall (u,v) \in V \times V$ instead of $\forall u \in V \quad \forall v \in V$? No ambiguity would occur, due to

Proposition 2.2. Assume $\vdash_Y \varphi(a,b)$ is defined whenever $\alpha: Y \to X$ and $a \in_Y A$, $b \in_Y B$. Then

$$\vdash_X \forall x \in A: (\forall y \in B: \varphi(x,y)) \tag{2.5}$$

if and only if

$$\vdash_X \forall z \in A \times B: \varphi(z) \tag{2.6}$$

Proof. Assume (2.5). Let $\alpha: Y \to X$ and let $c: Y \to A \times B$ be arbitrary; c is of form

$$(a,b): Y \longrightarrow A \times B$$

for $a \in_Y A$, $b \in_Y B$. By (2.5)

$$\vdash_Y \forall y \in B: \varphi(a,y), \tag{2.7}$$

and in particular

$$\vdash_Y \varphi(a,b),$$

that is, $\vdash_Y \varphi(c)$. This proves (2.6). Conversely, assume (2.6).

We must prove (2.7) for arbitrary $\alpha: Y \to X$ and $a: Y \to A$. So let $\beta: Z \to Y$ and $b: Z \to B$ be arbitrary. We must prove

$$\vdash_Z \varphi(a \circ \beta, b),$$

but this follows by applying (2.6) for the stage change $\alpha \circ \beta$ and the element $(a \circ \beta, b): Z \to A \times B$.

Since we have given a semantics for expressions of type $\vdash_X \varphi(a)$, it is clear what is meant by saying that one such expression can be deduced (= deduced semantically) from another. The proof of the Proposition given is an example of two such deductions. There is also the possibility of describing a notion of formal deduction, operating entirely with the $\vdash_X \varphi(a)$ as syntactic entitites; this we shall not go into.

EXERCISES

2.1. Prove that $\vdash_X \forall x \in R \;\; (\varphi(x) \Rightarrow \psi(x))$ if and only if for any $\alpha: Y \to X$ and any $a: Y \to R$, $\vdash_Y \varphi(a)$ implies $\vdash_Y \psi(a)$.

2.2. Let R_1 and R_2 be ring objects in E. Let $f: R_1 \to R_2$ be a map. Describe in terms of elements of R_1 and R_2 what it means for f to be a homomorphism of ring objects. Also, describe this by means of $\vdash$.

2.3. Let R_1 and R_2 be ring objects in E, and let B be an arbitrary object. Describe in terms of elements and $\vdash$ what it means for a map $f: B \times R_1 \to R_2$ to be a ring homomorphism with respect to the second variable.

2.4. Exercises 8.1 and 8.5 below may be solved now.

II.3: EXTENSIONS AND DESCRIPTIONS

Suppose $\varphi(x)$ is a mathematical formula about elements of an object $R \in \mathcal{E}$, for which we, for any $X \in \mathcal{E}$ and any element $a: X \to R$ have defined when

$$\vdash_X \varphi(a),$$

and assume φ is stable, i.e. for any $\alpha: Y \to X$, $\vdash_X \varphi(a)$ implies $\vdash_Y \varphi(a)$. An extension for φ is a monic map

$$e: F \rightarrowtail R$$

with $\vdash_F \varphi(e)$, and universal with this property, meaning: for any X and any $b: X \to R$,

$$\vdash_X \varphi(b) \text{ iff } b \text{ factors through } e.$$

(The implication '$\Leftarrow$' follows from stability of φ.) Clearly, if $e': F' \rightarrowtail R$ is another extension for φ, there is a unique map $f: F \to F'$ with $e' \circ f = e$, and this f is invertible; so the extension of φ is well defined as a subobject of R (cf. [55], V §7).

Assume that for each subobject of each object, we choose a representing monic map, which we denote $\hookrightarrow$ or $\subseteq$ rather than $\rightarrowtail$. (Frequently, such maps are omitted from notation.) Monic maps of form $\hookrightarrow$ are called inclusions. If the formula φ considered above has an extension, it has a unique extension which is an inclusion, and this inclusion is denoted

$$[\![\varphi]\!] \hookrightarrow R$$

or

$$[\![x \in R \mid \varphi(x)]\!] \hookrightarrow R.$$

Not all formulas need have extensions, but if E has finite inverse limits, and R is a ring object (say), any purely equational formula has one. For example, the D considered in §1 is the extension of the formula $x^2 = 0$.

If $C \rightarrowtail R$ is monic, and $b: X \to R$ is an arbitrary element of R, we define $\vdash_X b \in C$ to mean: b factors through $C \rightarrowtail R$. (Clearly, this is incomplete notation also, since the name of the map $C \to R$, rather than the name of its domain, should have been used.) So in particular, with φ as above,

$$\vdash_X b \in [\![x \in R \mid \varphi(x)]\!] \quad \text{iff} \vdash_X \varphi(b).$$

Also, we see that any inclusion $C \hookrightarrow R$ may be considered as an extension, namely of the formula $x \in C$.

We have the following

<u>Proposition 3.1</u> (Extensionality principle for subobjects). Assume $C_1 \hookrightarrow R$ and $C_2 \hookrightarrow R$ are subobjects. Then $C_1 \subseteq C_2$ (meaning $C_1 \hookrightarrow R$ factors through $C_2 \hookrightarrow R$) iff

$$\vdash_1 \forall x \in R: \ x \in C_1 \Rightarrow x \in C_2 \ .$$

<u>Proof</u>. The implication $\Rightarrow$ is evident. For $\Leftarrow$, use the assumption for the specific element $(C_1 \hookrightarrow R) \in_{C_1} R$.

Extensions for formulas φ which are defined only for elements b defined at stages later than a given stage X i.e. for configurations

$$\begin{array}{ccc} Y & \xrightarrow{\alpha} & X \\ & \searrow^{b} & \\ & & R \end{array}$$

also play a certain role; they require consideration of the comma category $\mathcal{E}/X$, cf. §6 below.

Suppose Φ is a mathematical law which to elements b of an object R_1 of $\mathcal{E}$ associates elements $\Phi(b)$ of an object $R_2 \in \mathcal{E}$, with $\Phi(b)$ being defined at the same stage as b, and assume Φ commutes with change-of-stage:

$$\alpha^*(\Phi(b)) = \Phi(\alpha^*(b))$$

for any $\alpha: Y \to X$, where X is the stage of definition of b. Then Φ is nothing but a natural transformation

$$\hom_{\mathcal{E}}(-,R_1) \longrightarrow \hom_{\mathcal{E}}(-,R_2),$$

and thus by Yoneda's lemma (see e.g. [55]. Corollary p.61) of form

$$\Phi(b) = f \circ b$$

for some unique $f: R_1 \longrightarrow R_2$. We say that f <u>is described by the law</u> Φ, or that Φ is a <u>description</u> of f. We shall refer to this principle for constructing maps f in $\mathcal{E}$ from laws Φ as the <u>Yoneda map construction principle</u>.

For example, if R is a ring object, the law

$$(a,b,d) \longmapsto a + d \cdot b$$

which to an element of $R \times R \times R$ associates an element of R, (using, for any given stage X the ring structure on the set $\hom_{\mathcal{E}}(X,R)$) describes a map $R \times R \times R \to R$. (Note that we implicitely use the correspondence (1.2) $\leftrightarrow$ (1.3), (for three factors).)

Maps described by laws which are only defined for elements b defined at stages later than a given stage X, i.e. laws which to configurations

$$\begin{array}{ccc} Y & \longrightarrow & X \\ & \searrow^{b} & \\ & & R_1 \end{array}$$

associate elements $\Phi(b) \in_Y R_2$, also play a role, and can be treated using the comma category $\mathcal{E}/X$, cf. §6. However, if $\mathcal{E}$ has products, such a law is clearly equivalent to a law which to elements of $X \times R_1$ associates elements of R_2, and thus describes a map

$$X \times R_1 \rightarrow R_2$$

which one may think of as an X-<u>parametrized family of maps from</u> R_1 to R_2.

The fact that the map described by a given description is unique can be formulated as

<u>Proposition 3.2</u> (Extensionality principle for maps). Assume f and g are maps $R_1 \longrightarrow R_2$. Then $f = g$ iff

$$\vdash_1 \forall x \in R_1 : \; f \circ x = g \circ x \qquad (3.3)$$

Thinking of f and g as laws that to elements of R_1 associate elements of R_2, and thus writing $f(x)$ instead of $f \circ x$, we may of course write (3.3) in the following way, which looks even more like the standard condition:

$$\vdash_1 \forall x \in R_1 : \; f(x) = g(x) \qquad (3.4)$$

The notations have been chosen so as to remind one as much as possible to the ones well tested in the category of sets, so that one is lead to expect certain properties well known from there. Such properties usually turn out to hold also in the general situation. We give four examples (Proposition 3.3-3.6); more follow in §4.

Consider arbitrary objects B and C in $\mathcal{E}$.

Proposition 3.3. Let $f: B \to C$ be a map. Then f is invertible iff

$$\vdash_1 \forall y \in C \ \exists! \, x \in B: \ f(x) = y,$$

(where $f(x)$, as in the remark after Proposition 3.2, denotes $f \circ x$).

Proof. The implication $\Rightarrow$ is trivial. For the other implication, consider the element $y = id_C \in_C C$. By assumption, there is a unique $x \in_C B$ with

$$f \circ x = y \quad (= id_C) \tag{3.5}$$

This x is a map $C \to B$, which is, by (3.5), a right inverse for f. On the other hand, by (3.5) we have

$$f \circ x \circ f = f = f \circ id_B.$$

Thus $x \circ f \in_B B$ and $id_B \in_B B$ are elements which satisfy

$$f \circ z_1 = f \circ z_2$$

when substituted for z_1 and z_2, respectively; such element is unique, by the assumption, hence $x \circ f = id_B$. So x is a two-sided inverse for f.

We may introduce a name, say g, for the inverse of f, in the case the conditions of Proposition 3.3 are satisfied. Then

$$\vdash_1 \forall y \in C \quad \forall x \in B\colon (g(y) = x) \Longleftrightarrow (y = f(x)).$$

There is a more general situation where names can be introduced. Suppose that B and C are arbitrary objects in $\mathcal{E}$, and that $\varphi(x,y)$ is a formula, where x and y range over elements of B and C, respectively.

<u>Proposition 3.4</u>. Assume $\vdash_1 \forall x \in B \ \exists !\, y \in C\colon \varphi(x,y)$. Then there exists a unique $g\colon B \longrightarrow C$ with

$$\vdash_1 \forall x \in B \quad \forall y \in C\colon \quad \varphi(x,y) \Longleftrightarrow y = g(x).$$

With the g thus defined,

$$\vdash_1 \forall x \in B\colon \quad \varphi(x,g(x)).$$

We omit the proof which is similar to that of Proposition 3.3.

What is the condition for satisfaction of a formula in which enters the symbol g, as introduced in this Proposition ? Assume $b \in_X B$, and let $\psi(y)$ be a formula with y ranging over elements of C, defined at stages later than X. Then, with φ and g as in Proposition 3.4, we have

<u>Proposition 3.5</u>. We have $\vdash_X \psi(g(b))$ if and only if

$$\vdash_X \exists !\, c \in C\colon \psi(c) \wedge \varphi(b,c) \tag{3.6}$$

<u>Proof</u>. Assume $\vdash_X \psi(g(b))$. Take $c = g(b)$; then certainly, $\vdash_X \psi(c)$; and $\vdash_X \varphi(b,c)$, since $\vdash_1 \forall x \in B\colon \varphi(x,g(x))$. Also, c is the unique element $\in_X C$ with this property, by the assumption of uniqueness in Proposition 3.4.

Conversely, assume $c\colon X \to C$ satisfies (3.6). Since $\vdash_X \varphi(b,c)$, and $\vdash_X \varphi(b,g(b))$, we get by uniqueness $c = g(b)$; by $\vdash_X \psi(c)$, we conclude $\vdash_X \psi(g(b))$.

For the final example, let φ_1 and φ_2 be formulae talking about elements of objects R_1 and R_2, respectively. Suppose they both have extensions:

$$H_1 = [\![x \in R_1 \mid \varphi_1(x)]\!] \hookrightarrow R_1$$

and

$$H_2 = [\![x \in R_2 \mid \varphi_2(x)]\!] \hookrightarrow R_2.$$

Then we have

Proposition 3.6. To describe a map $H_1 \longrightarrow H_2$, it suffices to describe a map $f\colon R_1 \to R_2$, with description $r \longmapsto \Phi(r)$, say, such that

$$\vdash_{\mathbf{1}} \forall x \in R_1\colon \varphi_1(x_1) \Rightarrow \varphi_2(\Phi(x)).$$

(Then the desired map $H_1 \to H_2$ is the restriction of f to H_1.)

EXERCISES

3.1. Prove that a map $f\colon R_1 \longrightarrow R_2$ is monic iff

$$\vdash_{\mathbf{1}} \forall x,y \in R_1\colon (f(x) = f(y)) \Rightarrow (x=y)$$

3.2. Let G be a monoid object in $\mathcal{E}$ (i.e. there is an associative $G \times G \longrightarrow G$ with a two sided unit $e\colon \mathbf{1} \to G$). Prove that G is a group object (i.e. construct an inversion map $G \to G$) if and only if

$$\vdash_{\mathbf{1}} \forall x \in G \quad \exists!\, y \in G\colon x \cdot y = e \wedge y \cdot x = e.$$

We assume now that E is a cartesian closed category. For any pair of objects, R and D, we have the exponential object R^D, and the bijective correspondence ("λ-conversion") for any $X \in E$

$$\frac{X \longrightarrow R^D}{X \times D \longrightarrow R} .$$

In particular, taking $X = R^D$, and taking the top map to be the identity of R^D, we get for the bottom map a map, denoted ev (for 'evaluation'),

$$R^D \times D \xrightarrow{\text{ev}} R,$$

(the 'end adjunction for the exponential adjointness').

Now let $f: X \to R^D$ and $d: X \to D$ be elements of R^D and D, respectively, defined at same stage. We define a certain element $f(d)$ of R (defined at stage X) by

$$f(d) := (X \xrightarrow{(f,d)} R^D \times D \xrightarrow{\text{ev}} R). \tag{4.1}$$

Note that ev has description

$$(f,d) \longmapsto f(d).$$

We have to face that we in Proposition 3.2 also introduced the notation $f(x)$, there with another meaning, namely $f \circ x$. This double use of notation is, by experience, known not to cause serious confusion. The situation where both meanings of the notation occur simultaneously is the following: there is given

$$\begin{array}{l} Y \xrightarrow{x} X \xrightarrow{f} R^D \\ \quad \searrow^{d} \\ \qquad\qquad D \, . \end{array} \tag{4.2}$$

We then have the element

$$Y \xrightarrow{(f \circ x, d)} R^D \times D \xrightarrow{ev} R, \tag{4.3}$$

which, by (4.1), is denoted $(f \circ x)(d)$. Interpreting x as an element of X (defined at stage Y), and utilizing the notation of Proposition 3.2, i.e. writing $f(x)$ for $f \circ x$, we arrive then at the notation $f(x)(d)$ for (4.3). Finally, interpreting $x: Y \to X$ as a change of stage from stage X to stage Y, and omitting it from notation, we have that $f \circ x$ is denoted just f, and so (4.3) is denoted $f(d)$; summarizing

$$(f \circ x)(d) = f(x)(d) = f(d). \tag{4.4}$$

The exponential adjoint $f^\vee: X \times D \to R$ of $f: X \to R^D$ may be described as the composite

$$X \times D \xrightarrow{f \times D} R^D \times D \xrightarrow{ev} R .$$

So, for $(x,d) \in_Y X \times D$ (i.e. for the situation (4.2)), we have

$$f^\vee(x,d) = ev \circ (f \circ x, d) = (f \circ x)(d) = f(x)(d).$$

This result, i.e.

$$f^\vee(x,d) = f(x)(d) \tag{4.5}$$

justifies the double use of the $f(\cdot)$ notation, because it is the standard way of rewriting a function in two variables x and d

into a function in one variable x whose values are functions in the other variable d, and vice versa (i.e., λ-conversion).

The third notation $f(d)$ occurring in (4.4) is essential in connection with the $\vdash$-notion. For instance, we have:

Proposition 4.1 (Extensionality principle for elements of function objects). Let $f_i \in_X R^D$, $(i = 1,2)$. Then

$$\vdash_X \forall d \in D: f_1(d) = f_2(d)$$

implies

$$\vdash_X f_1 = f_2.$$

(The converse implication is trivially true.)

Proof. It suffices to see $f_1^{\vee} = f_2^{\vee}: X \times D \to R$. By the extensionality principle for maps (Proposition 3.2), it suffices to see $f_1^{\vee}(x,d) = f_2^{\vee}(x,d)$ for an arbitrary pair (x,d) like in (4.2). But $f_1^{\vee}(x,d) = f_1(d)$, by (4.5) and (4.4), and similarly for f_2. The result now follows from the assumption $\vdash_X \forall d: f_1(d) = f_2(d)$, by considering the change of stage $x: Y \to X$.

How does one describe maps into function objects like R^D? To describe a map $f: X \to R^D$ is equivalent, by exponential adjointness, to describing a map $f^{\vee}: X \times D \longrightarrow R$. If $f^{\vee}$ is described by a law Φ which to an element

$$(x,d) \in_Y X \times D$$

associates an element

$$\Phi(x,d) \in_Y R$$

(at same stage), then we agree to use the following notation to describe f itself:

$$x \longmapsto [d \longmapsto \Phi(x,d)].$$

In other words, the conversion

$$\frac{X \times D \longrightarrow R}{X \longrightarrow R^D}$$

looks as follows in terms of descriptions

$$\frac{(x,d) \longmapsto \Phi(x,d)}{x \longmapsto [d \mapsto \Phi(x,d)]}. \tag{4.6}$$

We note that if $f\colon X \to R^D$ has description $x \longmapsto [d \longmapsto \Phi(x,d)]$, and we have given a d of D defined at a later stage $x\colon Y \to X$, i.e. we have a configuration (4.2), then (4.6) and (4.5) imply

$$f(x)(d) = \Phi(x,d) \in_Y R \tag{4.7}$$

If A and B are objects with some algebraic structure, say group objects, in a cartesian closed category with finite inverse limits, it is well known (cf. Exercise 4.2) that one out of B^A by an equalizer can carve 'the subobject $\underset{\sim}{\mathrm{Hom}}_{Gr}(A,B) \subseteq B^A$ of group homomorphisms', (and similarly for ring objects, module objects, etc.). We shall now describe these constructions using the technique introduced.

Let $f \in_X B^A$. Then

$$\vdash_X \; f \in \underset{\sim}{\mathrm{Hom}}_{Gr}(A,B)$$

if and only if

$$\vdash_X \; \forall (a_1,a_2) \in A \times A\colon \; f(a_1 \cdot a_2) = f(a_1) \cdot f(a_2) \tag{4.8}$$

If A and B are abelian group objects, and R is a ring object acting on A and B, making them into R-modules, then we can form a subobject of $\underset{\sim}{\mathrm{Hom}}_{\mathrm{Gr}}(A,B)$, denoted $\underset{\sim}{\mathrm{Hom}}_{R\text{-mod}}(A,B)$, having the property

$$\vdash_X f \in \underset{\sim}{\mathrm{Hom}}_{R\text{-mod}}(A,B)$$

if and only if

$$\vdash_X (f \in \underset{\sim}{\mathrm{Hom}}_{\mathrm{Gr}}(A,B), \text{ and } \forall r \in R \ \ \forall a \in A\colon f(r\cdot a) = r\cdot f(a)). \tag{4.9}$$

Similarly, if R is a ring object, and $R \to C_1$, $R \to C_2$ are R-algebra objects, we may form $\underset{\sim}{\mathrm{Hom}}_{R\text{-Alg}}(C_1,C_2)$, and give an elementwise description of this object in the style of (4.8) or (4.9).

We note that the description (4.8) of $\underset{\sim}{\mathrm{Hom}}_{\mathrm{Gr}}(A,B)$ may be interpreted as saying: $\underset{\sim}{\mathrm{Hom}}_{\mathrm{Gr}}(A,B) \hookrightarrow B^A$ is the extension of the formula $\forall(a_1,a_2)\colon f(a_1\cdot a_2) = f(a)\cdot f(a_2)$ (which is a formula with one free variable, f, ranging over B^A), and similarly for the other $\underset{\sim}{\mathrm{Hom}}$-objects.

One typical use of the method of describing maps by the elementwise description of §3 is when we make $\underset{\sim}{\mathrm{Hom}}_{\mathrm{Gr}}(A,B)$ (for A and B abelian group objects, written additively) into an abelian group object, by defining the addition

$$\underset{\sim}{\mathrm{Hom}}_{\mathrm{Gr}}(A,B) \times \underset{\sim}{\mathrm{Hom}}_{\mathrm{Gr}}(A,B) \xrightarrow{\ +\ } \underset{\sim}{\mathrm{Hom}}_{\mathrm{Gr}}(A,B)$$

by the description

$$(f_1,f_2) \longmapsto [a \longmapsto f_1(a) + f_2(a)] \quad .$$

By the convention used in (4.6), this describes first only a map

$$B^A \times B^A \xrightarrow{+} B^A \quad ;$$

we want to apply Proposition 3.6 to get the desired map. We should thus prove

$$\vdash_1 \forall f_1, f_2 \in B^A \times B^A : f_1 \text{ and } f_2 \text{ are homomprphisms}$$

$$\Rightarrow f_1 + f_2 \text{ is a homomorphism.}$$

This is a straightforward exercise using (4.6). Also Exercise 2.1 should be used, to shorten the proof.

EXERCISES

4.1. Let R be a ring object in a cartesian closed category E. Describe (using elements and descriptions as in §§ 1,3,4) a ring structure on R^B (where B is an arbitrary object).

If $B \to C$ is a map, prove that the induced map

$$R^C \longrightarrow R^B$$

is a homomorphism of ring objects (cf. Exercise 2.2).

Prove also that the composite

$$\underset{\sim}{\mathrm{Hom}}_{\mathrm{Ring}}(R_1,R_2) \times R_1 \hookrightarrow R_2^{R_1} \times R_1 \xrightarrow{\mathrm{ev}} R_2$$

is a ring homomorphism with respect to the second variable (cf. Exercise 2.3).

4.2. Let A and B be abelian group objects. Consider the maps

$$B^A \xrightarrow{\Delta} B^A \times B^A$$

$$B^A \times B^A \xrightarrow{B^{\mathrm{proj}_1} \times B^{\mathrm{proj}_2}} (B^{A\times A} \times B^{A\times A}) \cong (B \times B)^{A\times A}$$

$$(B \times B)^{A\times A} \xrightarrow{(+)^{A\times A}} B^{A\times A}$$

Prove that the composite of these three maps is a map $B^A \longrightarrow B^{A\times A}$ with description

$$f \longmapsto [(a_1,a_2) \longmapsto f(a_1) + f(a_2)] .$$

Similarly, construct in diagrammatic terms the map $B^A \longrightarrow B^{A\times A}$ with description

$$f \longmapsto [(a_1,a_2) \longmapsto f(a_1+a_2)]$$

(this is easier!). Argue that the equalizer of the two maps thus described is $\underset{\sim\sim}{\mathrm{Hom}}_{Ab}(A,B)$.

In a similar vein, if A and B are equipped with actions $\mu: R\times A \longrightarrow A$ and $\mu: R\times B \longrightarrow B$ of an object R, construct in diagrammatic terms two maps $B^A \longrightarrow B^{R\times A}$ whose equalizer is the extension of the formula

$$\forall r\in R \quad \forall a\in A: \ \mu(r,f(a)) = f(\mu(r,a))$$

(whose only free variable is f ranging over B^A).

4.3. Let $\sigma: R_1 \longrightarrow R_2^D$ (R_1, R_2, and D arbitrary objects). Prove that σ is monic iff

$$\vdash_1 \forall x,y\in R_1: (\forall d\in D: \sigma(x)(d) = \sigma(y)(d)) \Rightarrow x = y$$

(Hint: use Exercise 3.1 and Proposition 4.1.)

II.5: AXIOM 1 REVISITED

Basic in the diagrammatic (= variable free) formulation of Axiom 1 is the map

$$R \times R \xrightarrow{\alpha} R^D \qquad (5.1)$$

with description

$$(a,b) \longmapsto [d \longmapsto a + d \cdot b] \quad . \qquad (5.2)$$

We want to prove that the diagrammatic form of Axiom 1 ("α is invertible") is equivalent to the naive form ("every map $D \to R$ is uniquely of form $d \mapsto a + d \cdot b$"), provided we read the latter as talking about (generalized) elements of R^D, D, and R, as in the previous four paragraphs. To be precise, assume R is a commutative ring object in a cartesian closed category $\mathcal{E}$ with finite inverse limits, and define D as the extension

$$D = [\![x \in R \mid x^2 = 0]\!] \; .$$

Then we have

<u>Proposition 5.1</u>. The map α in (5.1) is invertible if and only if

$$\vdash_1 \forall f \in R^D \;\; \exists! (a,b) \in R \times R : \;\; \forall d \; f(d) = a + d \cdot b \qquad (5.3)$$

Proof. By Proposition 3.3, α is invertible iff

$$\vdash_1 \forall f \in R^D \quad \exists!(a,b) \in R \times R: \quad \alpha(a,b) = f.$$

Thus, it suffices to prove, for arbitrary elements f of R^D and (a,b) of $R \times R$, defined at same stage $(X$, say), that

$$\vdash_X \alpha(a,b) = f \quad \text{iff} \quad \vdash_X \forall d \in D: f(d) = a + d \cdot b.$$

Now, by the extensionality principle (Proposition 4.1)

$$\vdash_X \alpha(a,b) = f \quad \text{iff} \quad \vdash_X \forall d \in D: \alpha(a,b)(d) = f(d).$$

But $\alpha(a,b)(d) = a + d \cdot b$, by (4.7) and (5.2).

It is easy to deduce from (5.3) the other elementwise form of Axiom 1:

$$\vdash_1 \forall f \in R^D \ \exists!\ b \in R: \forall d \in D \quad f(d) = f(0) + d \cdot b \tag{5.4}$$

and vice versa. Combining (5.4) with Proposition 3.4 produces exactly the map $\gamma: R^D \to R$ ("principal part formation") of I.(1.3).

We now revisit some of the differential calculus which is based directly on Axiom 1.

First, if $U \hookrightarrow R$ is a subobject satisfying I.(2.3), that is, $\vdash_1 \forall x \in U, \forall d \in D: x+d \in U$, then one can deduce from (5.3) or (5.4) that

$$\vdash_1 \ \forall f \in R^U \ \forall x \in U \ \exists!\ b \in R: \ \forall d \in D \quad f(x+d) = f(x) + d \cdot b \tag{5.5}$$

By Proposition 3.4, we therefore have a unique map

$$R^U \times U \longrightarrow R \tag{5.6}$$

whose value at $(f,x) \in_X R^U \times U$ we denote $f'(x)$; $f'(x) \in_X R$. So by Proposition 3.4

$$\vdash_1 \quad \forall f \in R^U \quad \forall x \in U \quad \forall b \in R:$$

$$(\forall d \in D: f(x+d) = f(x) + d \cdot b) \iff (b = f'(x)),$$

and, as argued generally in Proposition 3.4, we then also have

$$\vdash_1 \quad \forall f \in R^U \quad \forall x \in U: \quad \forall d \; f(x+d) = f(x) + d \cdot f'(x), \qquad (5.7)$$

which is Taylor's formula, Theorem I.2.1.

EXERCISES

5.1. Prove that Axiom 1 can be formulated in a way which neither uses the cartesian closed structure of $\mathcal{E}$, nor any of the categorical logic introduced, as follows:

"For any $X \in \mathcal{E}$ and any $f: X \times D \to R$, there exists a unique $(a,b): X \to R \times R$ such that for any $\beta: Y \to X$ and any $d: Y \to D$, we have

$$f \circ (\beta, d) = a \circ \beta + d \cdot (b \circ \beta)$$

in $\hom_{\mathcal{E}}(Y,R)$".

5.2. Prove that Axiom 1 implies

$$\vdash_1 \forall x,y \in R: (\forall d \in D: d \cdot x = d \cdot y) \Rightarrow x = y.$$

II.6: COMMA CATEGORIES

Recall that if $\mathcal{E}$ is a category and $X \in \mathcal{E}$, then the category $\mathcal{E}/X$ (or $(\mathcal{E} \downarrow X)$, [55] II.6) of objects over X has for its objects the arrows in $\mathcal{E}$ with codomain X, and for morphisms commutative triangles

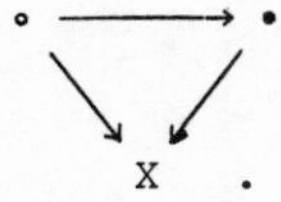

$\mathcal{E}/X$ always has a terminal object, namely $id_X: X \longrightarrow X$. If $\mathcal{E}$ has a terminal object $\mathbf{1}$, then $\mathcal{E} \cong \mathcal{E}/\mathbf{1}$ (letting $R \in \mathcal{E}$ correspond to $R \to \mathbf{1}$, the unique such map).

We henceforth assume that $\mathcal{E}$ has finite inverse limits. By making a choice of pull-back diagrams, we can to every $\alpha: Y \longrightarrow X$ associate a functor α^* ("pull-back functor),

$$\alpha^*: \mathcal{E}/X \longrightarrow \mathcal{E}/Y ,$$

which to an object $f \in \mathcal{E}/X$ associates that arrow which sits opposite to f in the chosen pull-back diagram

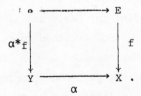

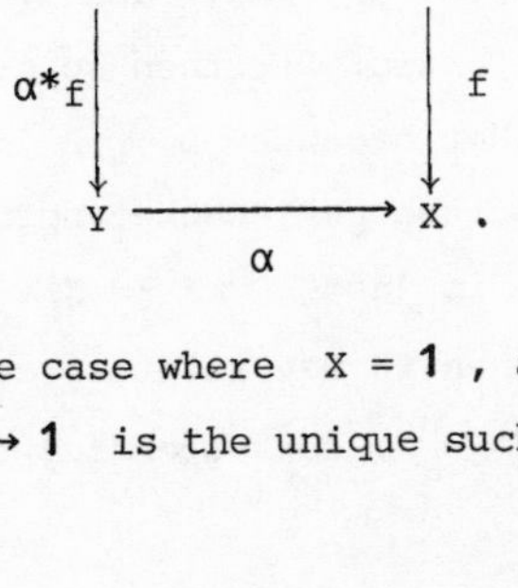

For the case where $X = \mathbf{1}$, and $\mathcal{E}/\mathbf{1}$ is identified with $\mathcal{E}$, and $\alpha: Y \longrightarrow \mathbf{1}$ is the unique such map, the functor $\alpha^*: \mathcal{E} \longrightarrow \mathcal{E}/Y$

can be described, on objects, by

$$R \longmapsto \begin{array}{c} R \times Y \\ \downarrow \mathrm{proj}_2 \\ Y \end{array} \qquad (6.1)$$

for $R \in \mathcal{E}$.

It is well known that, for any X, $\mathcal{E}/X$ has finite inverse limits (constructed in terms of those of $\mathcal{E}$), and that the pull-back functors $\alpha^*: \mathcal{E}/X \longrightarrow \mathcal{E}/Y$ preserve these limits.

Now, if $\alpha: Y \longrightarrow X$ is an object of $\mathcal{E}/X$, then there is a canonical isomorphism of categories

$$(\mathcal{E}/X)/\alpha \cong \mathcal{E}/Y \qquad (6.2)$$

Thus, any result about functors of the type (6.1) which holds for arbitrary $\mathcal{E}$ with finite inverse limits, holds for any pull-back functor α^*.

Also, if not only $\mathcal{E}$, but also all the $\mathcal{E}/X$, are cartesian closed (in which case we say that $\mathcal{E}$ is <u>stably</u> cartesian closed), the pull-back functors α^* preserve exponential objects,

$$\alpha^*(R^D) \cong \alpha^*(R)^{\alpha^*(D)}$$

for any $R, D \in \mathcal{E}/X$ and $\alpha: Y \to X$.

From these facts follows in particular that if R is a ring object in $\mathcal{E}$, then $R \times X \to X$ is a ring object in $\mathcal{E}/X$, and if R satisfies Axiom 1 in $\mathcal{E}$ (or Axiom 1',...,Axiom 2), then so does $R \times X \to X$ as a ring object in $\mathcal{E}/X$ (note that because $D \rightarrowtail R$ is formed by means of finite inverse limits and the pull-back functor preserves such, the D-object for $R \times X \to X$ is just $D \times X \longrightarrow X$).

Note also that if $f \in_X R$ in $\mathcal{E}$, then we may reinterpret f as a <u>global</u> element of the object $R \times X \longrightarrow X$ corresponding to

R in $\mathcal{E}/X$:

$$\begin{array}{ccc} X & \xrightarrow{(f,\mathrm{id}_X)} & R\times X \\ & {\scriptstyle \mathrm{id}_X}\searrow \quad \swarrow{\scriptstyle \mathrm{proj}_2} & \\ & X & \end{array}$$

and conversely.

This means, as a general principle, that if one wants to study some properties of generalized elements (of R, say), and the properties and the notions entering are preserved by all pull-back functors ("stable properties and notions"), then it suffices to study the property for global elements, (but in general categories).

For instance, if one wants to prove, say

$$\vdash_1 \forall f,g \in R^R \times R^R\colon\ (f\cdot g)' = f'\cdot g + f\cdot g', \tag{6.2}$$

this means proving something for arbitrary generalized elements $f,g\colon X \to R^R$, but since Axiom 1, and also the differentiation process (as can readily be proved) are stable, it suffices to prove (6.2) for global elements $f,g\colon 1 \longrightarrow R^R$, or by exponential adjointness, for $f,g\colon R \to R$ actual maps; however, now the proof should be valid in any model $\mathcal{E}', R'$ for Axiom 1, since we want to apply it for the case $\mathcal{E}' = \mathcal{E}/X$.

Under this viewpoint, for putting $\vdash_1$ in front of a sentence, it suffices that the sentence (the notions in it) are stable and satisfied for global elements in arbitrary models for the axiom.

It is clear from the preceding that the pull-back functors $\alpha^*\colon \mathcal{E}/X \to \mathcal{E}/Y$ are closely related to "change of stage" along α, as considered in §1, and, as there, it is a well tested abuse of notation to omit the symbol α^*. Thus, if we consider a ring object R in $\mathcal{E}$, the ring object

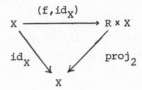

$$R \times X \xrightarrow{\text{proj}_2} X \quad \text{in } \mathcal{E}/X \tag{6.3}$$

may also just be denoted R. If $E \to X$ is an object in $\mathcal{E}/X$ which is equipped with structure of module over the ring object (6.3), this abuse of notation allows us to express this by just saying: it is an R-module object in $\mathcal{E}/X$.

For instance, if M is infinitesimally linear, the tangent bundle $TM \to M$ in $\mathcal{E}/M$ is an R-module object there.

If we think of an $\alpha: Y \to X$ as an element of X (defined at stage Y) rather than as a change of stage, and $(E \xrightarrow{f} X) \in \mathcal{E}/X$, then we would call $\alpha^*(f)$ (or its domain object in $\mathcal{E}$) for E_α, "the <u>fibre of</u> E <u>over the element</u> α".

We note that $E \to M$ gives rise to a <u>family</u> of objects E_m (= m^*E) <u>indexed</u> by the generalized elements $m: Y \to M$ of M.

It is immediate to analyze that, for $(E \to M)$ and $(F \to M)$ objects of $\mathcal{E}/M$, an element, defined at stage $m: Y \to M$, of $(F \to M)^{(E \to M)}$ is the same thing as (corresponds naturally to) a map in $\mathcal{E}$

$$E_m \longrightarrow F$$

commuting with the given maps to M. In particular, if $F = (R \times M \longrightarrow M)$, such maps, in turn, correspond bijectively to maps $E_m \to R$ in $\mathcal{E}$.

The way to interprete those formulas in the naive approach, where one variable ranges over objects <u>indexed</u> by a previous variable m (ranging over M, say), is by 'comprehending' the M-indexed family of objects into one object $E \to M$ in $\mathcal{E}/M$, and then use the previously introduced semantics, but now applied in the category $\mathcal{E}/M$. An example is Theorem I.13.2, and we shall by way of example refer to the theory of §I.13 on order and integration. We assume that R is a ring object in a stably cartesian closed category $\mathcal{E}$ with finite inverse limits, and that there is given a subobject

$$\textcircled{\le} \hookrightarrow R \times R;$$

we assume that elements of R are preordered by means of this, that is, for $x \in_X R$ and $y \in_X R$, we put

$$\vdash_X x \leq y \quad \text{iff} \quad (x,y) \text{ factors through } \textcircled{$\leq$}.$$

Of course, then, $\textcircled{$\leq$}$ is the extension $[\![(x,y) \in R \times R \mid x \leq y]\!]$. We assume that the relation $\leq$ satisfies the hypotheses of §I.13, like $\vdash_1 \forall (x,y,z) \in R \times R \times R: x \leq y \wedge y \leq z \Rightarrow x \leq z$, etc.

If b is a global element, $b: \mathbf{1} \longrightarrow R$, it makes sense to ask, for any $Y \in \mathcal{E}$ and any $a \in_Y R$, whether

$$\vdash_Y a \leq b,$$

and in fact, there is an extension for the formula $x \leq b$, namely the top composite in

$$\begin{array}{ccccc} [\![x \in R \mid x \leq b]\!] & \longrightarrow & \textcircled{$\leq$} & \xrightarrow{\mathrm{proj}_1} & R \\ \downarrow & & \downarrow{\scriptstyle \mathrm{proj}_2} & & \\ X & \xrightarrow[b]{} & R & & \end{array} \tag{6.4}$$

where $X = \mathbf{1}$ and where the square is formed as a pull-back.

Now, if X is not necessarily $\mathbf{1}$, the top composite is not monic in general, so it does not define a subobject $[\![x \in R \mid x \leq b]\!]$ of R in the category $\mathcal{E}$. However, the top composite, together with the displayed map $[\![x \in R \mid x \leq b]\!] \longrightarrow X$, define a monic map into $R \times X$, and the triangle

$$\begin{array}{ccc} [\![x \in R \mid x \leq b]\!] & \rightarrowtail & R \times X \\ & \searrow & \downarrow{\scriptstyle \mathrm{proj}_2} \\ & & X \end{array}$$

is then a subobject of $R \times X \to X$ in $\mathcal{E}/X$, or, with the abuse of notation introduced, a subobject of R in $\mathcal{E}/X$.

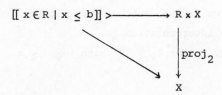

(We would have arrived at the same subobject by considering b as a global element of R in $\mathcal{E}/X$ and applying the construction of extension for the formula $x \leq b$ for this case.)

We conclude that the object $proj_2 : (\leq) \longrightarrow R$ in $\mathcal{E}/R$ is the comprehension of the family of objects $[\![x \in R \mid x \leq b]\!]$, where b ranges over elements of R; for, by the pull-back diagram in (6.4) above, it is seen to have the desired fibre for any $b \in_X R$ (X arbitrary). We denote this object in $\mathcal{E}/R$ by the symbol $[\![x \in R \mid x \leq b]\!]_{b \in R}$.

Similarly, we may form an object in $\mathcal{E}/R \times R$ (respectively in $\mathcal{E}/(\leq)$) which is the comprehension of the family of objects $[\![x \in R \mid a \leq x \leq b]\!]$, where (a,b) ranges over elements of $R \times R$ (respectively, over elements (a,b) of $R \times R$ with $a \leq b$). They are denoted similarly

$$[\![x \in R \mid a \leq x \leq b]\!]_{(a,b) \in R \times R} \quad \text{or} \quad [a,b]_{(a,b) \in R \times R}$$

and

$$[\![x \in R \mid a \leq x \leq b]\!]_{a \leq b} \quad \text{or} \quad [a,b]_{a \leq b} \tag{6.5}$$

respectively.

Now, if $\mathcal{E}$ is stably cartesian closed, we may in $\mathcal{E}/(\leq)$ form the exponential object R^C where C is short for (6.5). Its fibre over $(a,b) \in_X (\leq)$ is just $R^{[a,b]}$ (since pull-back functors preserve exponentials), so that R^C in $\mathcal{E}/(\leq)$ comprehends the family of objects $R^{[a,b]}$ indexed by all elements (a,b) of $(\leq)$.

This is the prerequisite for being able to interpret categorically the integration Theorem I.13.2, which involves this family of objects. The complete interpretation is:

"For any $X \in \mathcal{E}$, and any $(a,b) \in_X R \times R$ with $\vdash_X a \leq b$, we have in $\mathcal{E}/X$

$$\vdash_1 \forall f \in R^{[a,b]} \; \exists! \; g \in R^{[a,b]} : g(a) = 0 \wedge g' = f. \text{"} \tag{6.6}$$

Note that for given a,b as here, $R^{[a,b]}$ is a definite object in $\mathcal{E}/X$, so that (6.6) can be interpreted by applying the semantics of §2 for the category $\mathcal{E}/X$.

Another example where objects indexed by elements of another object is considered is in §I.6, where we consider the monads $\mathcal{M}_k(\underline{x})$ for $\underline{x} \in R^n$. The exposition given there is so as to immediately describe this family of objects in comprehended form, namely as the object I.(6.7) in $\mathcal{E}/R^n$. The complete interpretation of Corollary I.6.5 is then

"For any X and any $\underline{x} \in_X R^n$, we have in $\mathcal{E}/X$

$$\vdash_1 \forall f \in (R^m)^{\mathcal{M}_k(\underline{x})} \quad \forall \underline{z} \in \mathcal{M}_k(\underline{x}) : f(\underline{z}) \in \mathcal{M}_k(f(\underline{x}))\text{. "} \tag{6.7}$$

Note that, for given $\underline{x}$, all three objects $(R^m)^{\mathcal{M}_k(\underline{x})}$, $\mathcal{M}_k(\underline{x})$, and $\mathcal{M}_k(f(\underline{x}))$ are definite objects in $\mathcal{E}/X$ (where X is the stage of definition of $\underline{x}$), so that (6.7) can be interpreted by the semantics of §2 for the category $\mathcal{E}/X$.

Also, at some other places in Part I, indexed families of objects were considered, but always in terms of the comprehending object. Thus, the "individual tangent spaces T_xM for $x \in M$" are comprehended in (are the fibres of) the tangent bundle $TM \to M$. And in §I.17, we considered, for any formal manifold M, the "monads $\mathcal{M}_k(\underline{x})$ for $\underline{x} \in M$". They are comprehended by the object $M_{(k)} \longrightarrow M$ (and were in fact introduced in terms of this). In §§I.17 and I.18, we worked anyway, as the primary formulation, with diagrams, not requiring the semantics of §2 or the present §.

EXERCISES

6.1. Let E be a category with finite inverse limits. We say that an object $D \in E$ is exponentiable if $-\times D: E \to E$ has a right adjoint $(-)^D$. (So E is cartesian closed iff all its objects are exponentiable.) Prove that if f is exponentiable in E/X, then, for any $\alpha: Y \longrightarrow X$, $\alpha^*(f)$ is exponentiable in E/Y, and

$$\alpha^*(g^f) = \alpha^*(g)^{\alpha^*(f)} \quad .$$

II.7: DENSE CLASS OF GENERATORS

It is useful to have a more restricted notion of generalized element of an object R in a category $\mathcal{E}$, namely elements whose stage of definition belong to a suitable subclass. For instance, in the category Set of sets, it is well known that "it suffices" in some sense to consider only global elements, i.e. elements defined at stage $\mathbf{1}$.

Let $\mathcal{E}$ be a category and $\mathcal{A}$ a subclass of the class $|\mathcal{E}|$ of objects of $\mathcal{E}$. If $R \in \mathcal{E}$, an $\mathcal{A}$-element of R is a generalized element $r: Y \to R$ with $Y \in \mathcal{A}$. We may redefine the 'satisfaction'-notion of §2 by looking at $\mathcal{A}$-elements instead of arbitrary generalized elements.

Let us temporarily write $\vdash_{X,\mathcal{A}} \varphi$ to mean: for any $\alpha: Y \longrightarrow X$ with $Y \in \mathcal{A}$, we have $\vdash_Y \varphi$. Clearly for stable φ, $\vdash_X \varphi$ implies $\vdash_{X,\mathcal{A}} \varphi$, but we are interested in cases where this implication can be reversed.

The following definition can easily be seen to be equivalent to the classical definition, in [55] V.7, say.

Definition 7.1. We say that $\mathcal{A}$ is a class of generators if whenever f,g are maps with common domain (R_1, say), and with common codomain, then f = g iff

$$\vdash_{\mathbf{1},\mathcal{A}} \forall x \in R_1: f \circ x = g \circ x.$$

In the following, $\mathcal{A}$ is assumed to be a class of generators. Comparing then the Definition (which one might call: the $\mathcal{A}$-extensionality principle) with Proposition 3.2 (the extensionality principle), we see that for a formula of type $f \circ x = g \circ x$,

$\vdash_X$ and $\vdash_{X,A}$ are equivalent for any X. Also, it is easy to give an induction argument for equivalence of $\vdash_X$ and $\vdash_{X,A}$ for certain formulae:

Proposition 7.2. If φ_1 and φ_2 are stable formulae, for which, for any $\alpha\colon Y \to X$, $\vdash_Y \varphi_i$ iff $\vdash_{Y,A} \varphi_i$, $(i = 1,2)$, then

$$\vdash_X (\varphi_1 \Rightarrow \varphi_2) \quad \text{iff} \quad \vdash_{X,A} (\varphi_1 \Rightarrow \varphi_2), \tag{7.1}$$

and similarly for $\varphi_1 \wedge \varphi_2$ and $\forall x\, \varphi_1(x)$.

Proof. We do the case $\varphi_1 \Rightarrow \varphi_2$ only. The left hand side of (7.1) implies the right, because $\varphi_1 \Rightarrow \varphi_2$ is known to be stable. Conversely, assume $\vdash_{X,A} (\varphi_1 \Rightarrow \varphi_2)$. To prove $\vdash_X (\varphi_1 \Rightarrow \varphi_2)$, let $\alpha\colon Y \to X$ be such that $\vdash_Y \varphi_1$. For every $\beta\colon A \to Y$ with $A \in A$, we have $\vdash_A (\varphi_1 \Rightarrow \varphi_2)$ by assumption (applied to $A \to Y \to X$). Since φ_1 is stable, we have from $\vdash_Y \varphi_1$ also $\vdash_A \varphi_1$. So we conclude $\vdash_A \varphi_2$. Since this holds for every $\beta\colon A \to Y$, we conclude $\vdash_{Y,A} \varphi_2$. By assumption, then, $\vdash_Y \varphi_2$.

Note that in this proof, we did not use that A was a class of generators. Note also that we did not deal with formulae $\exists!\, x\, \varphi(x)$. This requires more on A, namely density, a classical category theoretic notion which we may formulate as follows: A is dense if the Yoneda map construction principle of §3 holds when instead of arbitrary generalized elements only A-elements are considered.

Proposition 7.3. Let A be a dense class of generators. Let $\varphi(x)$ be a stable formula about elements of an object R. Let X be an object such that for any $\alpha\colon Y \to X$ and $b \in_Y R$, we have $\vdash_Y \varphi(b)$ iff $\vdash_{Y,A} \varphi(b)$. Then

$$\vdash_X \exists!\, x\, \varphi(x) \quad \text{iff} \quad \vdash_{X,A} \exists!\, x\, \varphi(x).$$

Proof. The left hand side immediately implies the right, by stability. On the other hand, assume $\vdash_{X,A} \exists!\, x\ \varphi(x)$. For each $\alpha\colon A \to X$ with $A \in \mathcal{A}$, we thus have $\vdash_A \exists!\, x\ \varphi(x)$, so we have a unique element $b(\alpha) \in_A R$. So we have a law Φ which to an $\mathcal{A}$-element α of X associates an $\mathcal{A}$-element $b(\alpha)$ of R, and by the uniqueness, and stability of φ, we see that the law Φ is natural in $A \in \mathcal{A}$. So by the Yoneda map construction principle for $\mathcal{A}$-elements, $\Phi(\alpha) = f \circ \alpha$ for some (unique) $f\colon X \to R$. For this f, we have $\vdash_{X,\mathcal{A}} \varphi(f)$; for, let $\alpha\colon A \to X$ with $A \in \mathcal{A}$. Then $f \circ \alpha = \Phi(\alpha) =$ (the unique) $b\colon A \to R_1$ with $\vdash_A \varphi(b)$, so that $\vdash_A \varphi(f \circ \alpha)$. By the assumption, $\vdash_{X,\mathcal{A}} \varphi(f)$ implies $\vdash_X \varphi(f)$. The uniqueness of f follows because $\mathcal{A}$ is a class of generators.

Let us note that if C_1 and C_2 are subobjects of R, and $\mathcal{A}$ is dense, then the extensionality principle Proposition 3.1 holds even when only $\mathcal{A}$-elements are considered, because if every $\mathcal{A}$-element of C_1 is also an element of C_2, we can use the Yoneda map construction principle for $\mathcal{A}$-elements, to construct a map $C_1 \to C_2$.

EXERCISES

7.1. Prove that in the category Set, the class consisting of the only object $\mathbf{1}$ (= terminal object) is a dense class of generators.

7.2. Prove that a dense class $\mathcal{A}$ is automatically a class of generators, provided $\mathcal{E}$ has equalizers.

7.3. Assume $\mathcal{A} \subseteq |\mathcal{E}|$ is a dense class of generators, and let $\mathcal{A}$ also denote the full subcategory of $\mathcal{E}$ consisting of the objects from $\mathcal{A}$. Prove that the composite functor

$$\mathcal{E} \xrightarrow{\ y\ } \underline{\mathrm{Set}}^{\mathcal{E}^{op}} \xrightarrow{\ r\ } \underline{\mathrm{Set}}^{\mathcal{A}^{op}} \tag{7.2}$$

is full and faithful (where y is the Yoneda embedding, and where r is the restriction functor along the inclusion $A^{op} \hookrightarrow E^{op}$). Conversely, if (7.2) is full and faithful, the objects of A form a dense class of generators. (This is Isbell's <u>adequacy</u> notion for $A \hookrightarrow E$.)

7.4. If $\mathbb{A}$ is any category, the representable functors in $Set^{\mathbb{A}^{op}}$ form a dense class of generators.

7.5. Assume A is a dense class of generators in a cartesian closed category E, and that R is a commutative ring object in E. Strengthen Exercise 5.1 in the following way: prove that Axiom 1 can be formulated:

> "For any $X \in A$ and any $f: X \times D \to R$, there exists a unique $(a,b): X \to R \times R$ such that for any $\beta: Y \to X$ with $Y \in A$ and any $d: Y \to D$, we have
> $f \circ (\beta, d) = a \circ \beta + d \cdot (b \circ \beta)$
> in $\hom_E(Y,R)$."

II.8: SATISFACTION (2), AND TOPOLOGICAL DENSITY

The satisfaction relation $\vdash$ described in §2 dealt only with the logical constructs

$$\forall,\ \exists!\ \wedge,\ \text{and}\ \Rightarrow\ ,$$

and the whole of Part I except §21 was designed so that only these logical constructs were used. There does, however, exist synthetic considerations where the logical constructs

$$\exists,\ \vee,\ \text{and}\ \neg\ \text{(negation)}$$

are used, and some of these will occur in Part III (notably §10). To give a satisfaction relation $\vdash$ for these logical constructs is less elementary, and requires that the category E in which the interpretation takes place is equipped with the further structure of a Grothendieck topology, or equivalently that E is made into a site (see e.g. [28] 0.3 for these notions); we shall assume that E is equipped with a Grothendieck pretopology which is a slightly stronger notion, and we describe that explicitely in Appendix B. In particular, for each $X \in E$ we are given a class Cov(X) of families

$$\{\alpha_i : X_i \longrightarrow X \mid i \in I\} \qquad (8.1)$$

of 'coverings' of X.

We shall be interested in the case where there is given a dense class A of generators in E. A covering (8.1) will be called an A-covering if each $X_i \in A$. We shall assume furthermore that each X has an $\mathcal{A}$-covering ("$\mathcal{A}$ is topologically dense"). (These assumptions are satisfied for the case where $\mathcal{E} = \widetilde{\mathbb{C}}$ (= sheaves on a small category $\mathbb{C}$ equipped with a subcanonical Grothendieck topology, (see Appendix B)), with $\mathcal{A}$ = class of representables, $\{y(C) \mid C \in \mathbb{C}\}$.)

For $\mathcal{E}$ a category equipped with a Grothendieck pretopology and a dense class of generators, $\mathcal{A}$, which is topologically dense, we now describe the satisfaction relation for $\exists$, $\vee$, and negation, in continuation of §2. For the sake of illustration, we consider a fixed object R, as there.

Suppose $X \in \mathcal{E}$ and suppose $\varphi(x)$ is a mathematical formula for which we for any $\alpha: Y \to X$ and any element $b \in_Y R$ have already defined what we mean by

$$\vdash_Y \varphi(b) \quad . \tag{8.2}$$

Then we write

$$\vdash_X \exists x\ \varphi(x)$$

if there exists an $\mathcal{A}$-covering $\{\alpha_i : X_i \to X \mid i \in I\}$ such that, for each $i \in I$, there exists an element $b_i \in_{X_i} R$ with $\vdash_{X_i} \varphi(b_i)$.

Similarly, suppose $X \in \mathcal{E}$ and suppose φ and ψ are mathematical formulae for which we, for any $\alpha: Y \to X$ have already defined when

$$\vdash_Y \varphi \text{ and when } \vdash_Y \psi \quad .$$

Then we write

$$\vdash_X \varphi \vee \psi$$

if there exists an $\mathcal{A}$-covering $\{\alpha_i : X_i \to X \mid i \in I_1 \cup I_2\}$ such that

$$\vdash_{X_i} \varphi \quad \text{for all} \quad i \in I_1$$

and

$$\vdash_{X_i} \psi \quad \text{for all} \quad i \in I_2.$$

We can make an analogous description of

$$\vdash_X \bigvee_j \varphi_j$$

for arbitrary, even infinite disjunctions; we leave that to the reader. Finally, we write

$$\vdash_X \neg \varphi$$

if, whenever $\alpha: Y \to X$ is such that $\vdash_Y \varphi$, then the empty family is a covering of Y.

With the stability concept of § 2, it is possible to prove

Proposition 8.1. If φ and ψ are stable, then so are

$\exists\, x\, \varphi(x)$, $\varphi \vee \psi$, and $\neg\varphi$.

We omit the proof which is a straightforward consequence of the stability properties which a covering notion must satisfy to qualify as a Grothendieck pretopology; cf. [30] or [67].

Besides the stability concept for formulae, we have a localness-concept: we say that φ is local if for any A-covering $\{\alpha_i: X_i \to X \mid i \in I\}$,

$$(\vdash_{X_i} \varphi \quad \text{for all} \quad i \in I) \text{ implies } \vdash_X \varphi\,.$$

Proposition 8.2. If φ and ψ are local and stable, then so are

$$\forall x\, \varphi(x), \quad \exists x\, \varphi(x), \varphi \Rightarrow \psi, \varphi \wedge \psi, \quad \varphi \vee \psi, \text{ and } \quad \varphi.$$

Again, we refer to [30] or [67] for a proof.

It is also easy to see (using Propositions 7.2 and 7.3, and the fact that the composite of two coverings is a covering): if A_1 and A_2 are both a dense set of generators and topologically dense, then then resulting $\vdash$ notion is the same.

For Proposition 8.2 to be useful as an induction principle, we need to have some local and stable formulas to start with. For this, we need to add the further assumption that each covering family is jointly epic, because this immediately implies that the formula $x = y$ is local. For, if

$$X \underset{b}{\overset{a}{\rightrightarrows}} R$$

(with X and R arbitrary) are such that there is a covering $\{X_i \longrightarrow X \mid i \in I\}$ with $\vdash_{X_i} a = b$, then since the family is jointly epic, we conclude $a = b$, i.e. $\vdash_X a = b$.

Also, we have the problem to compare the $\exists!$-notion of §3 with the $\exists$-notion in the present §. This turns out to hinge exactly on the classical notion of the Grothendieck topology being subcanonical (cf. Appendix B), which implies also that covering families are jointly epic. We have

Proposition 8.3. Assume we have a subcanonical Grothendieck topology on E. Then the following two assertions are equivalent:

$$\vdash_X \exists!\, x\, \varphi(x)$$

and

$$\vdash_X \exists x\ \varphi(x) \wedge (\forall x,y\colon \varphi(x) \wedge \varphi(y) \Rightarrow (x=y)).$$

Also $\exists!\ x\ \varphi(x)$ is stable if φ is.

<u>EXERCISES</u> ("Propositional logic")

8.1. Prove that if $\vdash_X \varphi \Rightarrow \psi$ and $\vdash_X \psi \Rightarrow \theta$, then $\vdash_X \varphi \Rightarrow \theta$.

8.2. Assume that φ_1 and φ_2 are stable and ψ is local. Prove that if $\vdash_X \varphi_1 \Rightarrow \psi$ and $\vdash_X \varphi_2 \Rightarrow \psi$, then $\vdash_X (\varphi_1 \vee \varphi_2) \Rightarrow \psi$. (The converse is easy, by 8.1.)

8.3. Prove that $\vdash_X \neg(\varphi_1 \vee \varphi_2)$ is equivalent to $\vdash_X \neg\varphi_1 \wedge \neg\varphi_2$ (this may be seen as a special case of 8.2). This is one of the de Morgan rules. The other de Morgan rule $(\neg(\varphi_1 \wedge \varphi_2)$ iff $\neg\varphi_1 \vee \neg\varphi_2)$ does not hold in general.

8.4. Prove that $\vdash_X \varphi$ implies $\vdash_X \neg\neg\varphi$.

8.5. Prove that $\vdash_X \varphi \Rightarrow (\psi \Rightarrow \theta)$ iff $\vdash_X (\varphi \wedge \psi) \Rightarrow \theta$.

8.6. Prove that $\vdash_X \varphi \Rightarrow \neg\psi$ iff $\vdash_X \neg(\varphi \wedge \psi)$ (this may be seen as a special case of 8.5).

8.7. Prove that if $\vdash_X \varphi \Rightarrow \psi$, then $\vdash_X (\neg\psi) \Rightarrow (\neg\varphi)$.

8.8. Prove that

$$\vdash_X (\bigvee_i \varphi_i) \Rightarrow \neg(\bigwedge_j \psi_j) \qquad i = 1,\ldots,n, \quad j = 1,\ldots,m$$

if and only if

$$\vdash_X \bigwedge_i \neg(\varphi_i \wedge \bigwedge_j \psi_j),$$

and that the latter is implied (for $n = m$) by

$$\vdash_X \neg(\varphi_1 \wedge \psi_1) \text{ and } \ldots \text{ and } \vdash_X \neg(\varphi_n \wedge \psi_m)$$

and likewise, (for $n = m$), by

$$\vdash_X \bigwedge_{i=1}^{n} (\varphi_i \Rightarrow \neg\psi_i).$$

Hint: use Exercises 8.2 and 8.6.

II.9: GEOMETRIC THEORIES

We consider formulas which are built from polynomial equations with coefficients from k (an arbitrary commutative ring in Sets) by means of conjunctions $\wedge$, disjunctions $\bigvee$ (possible infinite), and $\exists$. Such are in [28] called geometric formulae (in the language of the theory of k-algebras).

Examples: The formula "x is invertible" (i.e. $\exists y: x \cdot y = 1$) is geometric; the formula "x is nilpotent" (i.e.

$$(x = 0) \vee (x^2 = 0) \vee (x^3 = 0) \vee \ldots \qquad ,$$

is geometric. Both these examples have one free variable, x.

A geometric sentence is a sentence (without free variables) of form

$$\forall x_1,\ldots,x_n: \varphi(x_1,\ldots,x_n) \Rightarrow \psi(x_1,\ldots,x_n) \tag{9.1}$$

where φ and ψ are geometric formulae in n free variables $x_1,\ldots,x_n$; n may be 0. Also

$$\forall x_1,\ldots,x_n: \neg\varphi(x_1,\ldots,x_n) \tag{9.2}$$

as well as

$$\forall x_1,\ldots,x_n: \psi(x_1,\ldots,x_n) \tag{9.3}$$

are counted as geometric sentences if φ and ψ are geometric formulae. (These may be seen as special cases of (9.1), by taking ψ to be 'false' and φ to be 'true', respectively, i.e. an empty disjunctions and an empty conjunction, respectively.)

A <u>geometric theory</u> (in the language of the theory of k-algebras) is one whose axioms are geometric sentences.

If the formulas are built not only from polynomial equations, but also from inequalities $\leq$ between polynomials, we get similarly the notion of geometric formula/sentence/theory in the language of the theory of preordered k-algebras.

Here are some examples of geometric sentences ("x is invertible" is short for "$\exists y: x \cdot y = 1$"):

$$\neg(0 = 1) \tag{9.5}$$

$$\forall x: \ (x \text{ is invertible}) \vee (1-x \text{ is invertible}) \tag{9.6}$$

$$\forall x: \ (x \text{ is invertible}) \vee (x = 0) \tag{9.7}$$

$$\forall x: \ (x \text{ is invertible}) \Rightarrow ((\exists z: z^2 = x) \vee (\exists z: z^2 = -x)) \tag{9.8}$$

in the language of the theory of $\mathbb{Z}$-algebras (commutative rings). An example of a geometric sentence in the language of preordered rings is

$$\forall x: ((x \leq 0) \vee (x \leq 1) \vee (x \leq 2) \vee \dots \quad). \tag{9.9}$$

An example of a sentence, which is not geometric, is

$$\forall x \ (\neg(x = 0) \Rightarrow x \text{ is invertible}), \tag{9.10}$$

more generally, the sentence occurring in III.(2.2) below.

The sentences (9.5) and (9.6) (together with the equational sentences that define the notion of commutative ring) are the axioms for the theory of <u>local</u> rings, which is thus a geometric theory. Similarly, (9.5) and (9.7) give a theory of fields which is geometric; whereas (9.10) describes a different, non-geometric, field notion. This non-geometric field notion is compatible with

Axiom 1, and will be studied in Part III, §10.

When we say that a ring object R in a category $\mathcal{E}$ with a Grothendieck topology, is a <u>local</u> ring object we of course mean that we have

$$\vdash_1 \neg(0 = 1) \tag{9.12}$$

and

$$\vdash_1 \forall x: (x \text{ is invertible}) \vee (1-x \text{ is invertible}) \tag{9.13}$$

(where the variables range over elements of R). Similarly for any other 1st order theory $\mathcal{T}$. We shall also say: "R is a <u>model</u> of $\mathcal{T}$". In the applications, $\mathcal{E}$ will be a topos, where there is an easily described canonical topology: a family $\{X_i \to X \mid i \in I\}$ is covering if it is jointly epic. This topology is subcanonical, i.e. the conclusion of Proposition 8.3 holds.

It is of course possible to describe the notion of 'local ring object' in a topos $\mathcal{E}$ without using the systematic approach of the semantics of §8. Thus, a ring object $R \in \mathcal{E}$ is local iff $0: \mathbf{1} \longrightarrow R$ and $1: \mathbf{1} \longrightarrow R$ have empty interesection (this is the semantics-free way of saying $\neg(0 = 1)$), and

$$\mathrm{Inv}(R) \cup (\mathrm{Inv}(R) + 1) = R,$$

where

$$\mathrm{Inv}(R) = [\![x \mid \exists! y: x \cdot y = 1]\!]$$

and

$$\begin{aligned} \mathrm{Inv}(R) + 1 &= [\![x \mid \exists! y: (x-1) \cdot y = 1]\!] \\ &= [\![x \mid x-1 \in \mathrm{Inv}(R)]\!] . \end{aligned} \tag{9.14}$$

(These objects can also be described diagrammatically, without using the concept of extension; cf. III §3 below.)

PART III

MODELS

INTRODUCTION

The question of models should be viewed from two angles. One is the purely synthetic: we consider some property or structure which we out of experience think the geometric line has, and we experiment with the property to see whether it is logically consistent with other desirable or true properties. To this end we construct models, (which are often of quite algebraic character).

The other angle from which we view the question of models is to compare the synthetic theory with the analytic, for the benefit of both. The mutual benefit may for instance take form of a definite comparison theorem to the effect that properties proved or constructions performed synthetically hold or exist in the analytic theory, too, or vice versa. The models that give rise to the comparisons usually contain the category of smooth manifolds as a full subcategory, and are called well-adapted models (= well adapted for the comparison).

In §§ 1-2 we treat algebraic models, and in the rest the well-adapted ones. The latter are treated by a quite algebraic method, namely using the "algebraic theory of smooth functions".

III.1: MODELS FOR AXIOMS 1, 2, AND 3

All models we present will be closely related to categories of form $\mathcal{E} = \underline{\text{Set}}^{\mathcal{R}}$, where $\mathcal{R}$ is some small category of rings, and with R the forgetful functor $\mathcal{R} \longrightarrow \underline{\text{Set}}$; it carries a canonical 'argumentwise' ring object structure, whose addition, say, $R \times R \xrightarrow{+} R$, has for its B-component $(B \in \mathcal{R})$

$$(R \times R)(B) \longrightarrow R(B)$$

the addition map $B \times B \longrightarrow B$ (note $(R \times R)(B) = R(B) \times R(B) = B \times B$, and $R(B) = B$).

We have the Yoneda embedding $y\colon \mathcal{R}^{op} \longrightarrow \underline{\text{Set}}^{\mathcal{R}}$, which is full and faithful, and the objects of form yB for $B \in \mathcal{R}$ (i.e. the representable functors) form a dense set of generators.

Let k be a commutative ring in $\underline{\text{Set}}$. We use the notation $FP\mathbb{T}_k$, $FG\mathbb{T}_k$, and $\mathbb{T}_k$-Alg, for the category of finitely presented, finitely generated, and all (commutative) k-algebras, respectively. If $\mathcal{E}$ is a category with finite inverse limits, $\mathbb{T}_k\text{-Alg}(\mathcal{E})$, denotes the category of k-algebra objects in $\mathcal{E}$. Finally, if $R \in \mathbb{T}_k\text{-Alg}(\mathcal{E})$, the comma-category $(R \downarrow \mathbb{T}_k\text{-Alg}(\mathcal{E}))$ of "objects under R" (cf. [55],II.6) is denoted R-Alg. All this is consistent with the usage in I §12. The reader acquainted with finitary algebraic theories (in the sense of Lawvere [49] ; cf. Appendix A) will see that in the present §, and in the next, $\mathbb{T}_k$ may be replaced by any other finitary algebraic theory $\mathbb{T}$, and in §8, this will be essential.

If $\mathcal{E}$ is cartesian closed with finite inverse limits, we may form the hom-object $\underset{\sim}{\text{Hom}}_{R\text{-Alg}}(C,C')$ for any $C,C' \in$ R-Alg (where $R \in \mathbb{T}_k\text{-Alg}(\mathcal{E})$), as in I §12, and we have the functor

$\mathrm{Spec}_C\colon (\mathrm{FP}\mathbb{T}_k)^{op} \longrightarrow \mathcal{E}$, as there. If further $\mathcal{E}$ is proper left exact (cf. Appendix A), Spec_C extends to a proper left exact functor $(\mathrm{FG}\mathbb{T}_k)^{op} \longrightarrow \mathcal{E}$, also denoted Spec_C.

Let $\mathcal{R}$ be a small full subcategory of the category $\mathbb{T}_k$-Alg. Let $\mathcal{E} \hookrightarrow \underline{\mathrm{Set}}^{\mathcal{R}}$ be a full subcategory such that

(i) all representables $y(B')$ for $B' \in \mathcal{R}$, are in $\mathcal{E}$,

(ii) the inclusion functor is proper left-exact, and preserves formation of exponential objects

(iii) the forgetful functor $\mathcal{R} \to \underline{\mathrm{Set}}$ belongs to $\mathcal{E}$.

(Example: $\mathcal{E} = \widetilde{\mathcal{R}^{op}}$ for some subcanonical topology on $\mathcal{R}^{op}$ for which R is a sheaf). Then $\mathrm{Spec}_R\colon (\mathrm{FG}\mathbb{T}_k)^{op} \longrightarrow \underline{\mathrm{Set}}^{\mathcal{R}}$, and $y\colon \mathcal{R}^{op} \longrightarrow \underline{\mathrm{Set}}^{\mathcal{R}}$, factor through $\mathcal{E}$, and because of (ii), the formation of $\underset{\sim}{\mathrm{Hom}}_{R\text{-Alg}}$ is preserved by the inclusion functor.

Note that, for fixed $B'' \in \mathcal{R}$, the functor "evaluation at B" ": $\underline{\mathrm{Set}}^{\mathcal{R}} \to \underline{\mathrm{Set}}$ preserves all inverse limits, and C being an R-algebra implies that $C(B'')$ is a B''-algebra; from this we conclude

$$\mathrm{Spec}_C(B)(B'') = \mathrm{Spec}_{C(B'')}(B).$$

We say that an object $B \in \mathrm{FG}\mathbb{T}_k$ is stable with respect to $\mathcal{R}$ if $B' \in \mathcal{R}$ implies $B' \otimes_k B \in \mathcal{R}$ (where $\otimes_k$ is the coproduct in $\mathbb{T}_k$-Alg). We have, with $\mathcal{E}$ and R as above:

Lemma 1.1. If B is stable with respect to $\mathcal{R}$, then for any $B' \in \mathcal{R}$,

$$y(B') \times \mathrm{Spec}_R(B) \cong y(B' \otimes_k B).$$

Proof. The values of the two sides on a fixed $B'' \in \mathcal{R}$ are, respectively,

$$\hom_{\mathcal{R}}(B',B'') \times \mathrm{Spec}_{B''}(B)$$

and

$$\hom_{k\text{-Alg}}(B' \otimes_k B,B'') \quad (\cong \hom_{k\text{-Alg}}(B',B'') \times \hom_{k\text{-Alg}}(B,B'')).$$

For fixed B' and B'', these expressions describe proper left exact functors in $B \in (FG\amalg_k)^{op}$. Also they have the same value on $B = F(n)$, since

$$\hom_{\mathcal{R}}(B',B'') \times (B'')^n \cong \hom_{k\text{-Alg}}(B',B'') \times \hom_{k\text{-Alg}}(F(n),B'')$$

$$\cong \hom_{k\text{-Alg}}(B' \otimes_k F(n),B'').$$

Naturality, in connection with the result of Appendix A, yields the result.

For any $C \in R\text{-Alg}$, and any $B \in FG\amalg_k$, we have a canonical map in $\mathcal{E}$

$$\underset{\sim}{\mathrm{Hom}}_{R\text{-Alg}}(R^{\mathrm{Spec}_R(B)},C) \xrightarrow{\nu_{B,C}} \mathrm{Spec}_C(B), \tag{1.1}$$

as in I.(12.2).

Theorem 1.2. For any $B \in FG\amalg_k$ which is stable with respect to $\mathcal{R}$, the map $\nu_{B,C}$ in (1.1) is an isomorphism for all $C \in R\text{-Alg}$.

Proof. Since all constructions involved are preserved by $\mathcal{E} \hookrightarrow \underline{\mathrm{Set}}^{\mathcal{R}}$, we may as well assume $\mathcal{E} = \underline{\mathrm{Set}}^{\mathcal{R}}$, so both sides in (1.1) are functors $\mathcal{R} \to \underline{\mathrm{Sets}}$. For each $B' \in \mathcal{R}$, we analyze the B'-component of the natural transformation $\nu_{B,C}$ as follows.

Using the universal property of Spec-functors, we easily see

$$\mathrm{Spec}_C(B)(B') = \mathrm{Spec}_{C(B')}(B) \cong \hom_{\amalg_k\text{-Alg}}(B, C(B')); \qquad (1.2)$$

on the other hand

$$\underset{\sim\sim}{\mathrm{Hom}}_{R\text{-Alg}}(R^{\mathrm{Spec}_R(B)}, C)(B') \cong \hom_{\mathcal{E}}(yB', \underset{\sim\sim}{\mathrm{Hom}}_{R\text{-Alg}}(R^{\mathrm{Spec}_R(B)}, C)),$$

by Yoneda's lemma; by exponential adjointness, this set in turn is

$\cong$ set of maps in $\mathcal{E}$

$$yB' \times R^{\mathrm{Spec}_R(R)} \xrightarrow{\ \tau\ } C$$

which are R-algebra homomorphisms in the 2nd variable

$\cong$ set of $\mathcal{R}$-indexed natural families of maps

$$(yB')(B'') \times R^{\mathrm{Spec}_R(B)}(B'') \xrightarrow{\ \tau_{B''}\ } C(B''),\ B'' \in \mathcal{R} \qquad (1.3)$$

which are B"-algebra homomorphisms in the 2nd variable.

Now, if B is stable with respect to $\mathcal{R}$,

$$R^{\mathrm{Spec}_R(B)}(B'') \cong \hom_{\mathcal{E}}(yB'', R^{\mathrm{Spec}_R(B)})$$

$$\cong \hom_{\mathcal{E}}(yB'' \times \mathrm{Spec}_R(B), R)$$

$$\cong \hom_{\mathcal{E}}(y(B'' \otimes_k B), R) \quad \text{(by Lemma 1.1)}$$

$$\cong B'' \otimes_k B,$$

so the data of the $\tau_{B''}$ (for $B'' \in \mathcal{R}$) can equivalently be described

$\cong$ set of $\mathcal{R}$-indexed natural families of maps

$$\hom_{\mathcal{R}}(B',B'') \times (B'' \otimes_k B) \xrightarrow{\tau_{B''}} C(B''), \qquad B'' \in \mathcal{R}$$

which are B''-algebra homomorphisms in the 2nd variable.

But the data: B''-algebra homomorphisms $B'' \otimes_k B \longrightarrow C(B'')$ is equivalent (by composing with the inclusion $B \to B'' \otimes_k B$) to: $\mathbb{T}_k$-algebra homomorphisms $B \longrightarrow C(B'')$, so that the data of the $\tau_{B''}$ can equivalently be described

$\cong$ set of $\mathcal{R}$-indexed natural families of maps

$$\hom_{\mathcal{R}}(B',B'') \times B \xrightarrow{\tau_{B''}} C(B''), \qquad B'' \in \mathcal{R} \tag{1.4}$$

which are $\mathbb{T}_k$-algebra homomorphisms in the 2nd variable.

Out of this data, we can get a $\mathbb{T}_k$-algebra map $t: B \to C(B')$, namely $t(b) := \tau_{B'}(\mathrm{id}_{B'},b)$. On the other hand, given a $\mathbb{T}_k$-algebra map $t: B \to C(B')$, we construct a family (1.4) by putting, for $\beta: B' \to B''$ and $b \in B$,

$$\tau_{B''}(\beta,b) := C(\beta)(t(b)),$$

and it is immediate to check that these two processes are mutually inverse.

The fact that the passage from the set (1.2) to the set (1.4) really comes about from the $\nu_{B,C}$ (which was given an independent description) is left to the reader, who may get help from Exercises 1.1-1.4.

<u>Corollary 1.3</u>. Let $\mathcal{R} = \mathrm{FP}\mathbb{T}_k$ (= finitely presented k-algebras). Then $\mathcal{E} = \mathrm{Set}^{\mathcal{R}}$, with R = forgetful functor $\mathcal{R} \to$ <u>Sets</u>, satisfies Axiom 2_k, and in particular Axiom 1_k^W.

Proof. Since $FP\amalg_k$ is closed under the formation of $\otimes_k$-products in $\amalg_k$-Alg, the result follows immediately from the Theorem.

The same result holds with $FG\amalg_k$ instead of $FP\amalg_k$, or with the category of all k-algebras of cardinality $< \alpha$ for some suitable large cardinal number α (this is essentially the category studied in [9]), but $\mathcal{R} = FP\amalg_k$ has the nice feature of $\underline{Set}^{\mathcal{R}}$ being the classifying topos for the theory of (commutative) k-algebras, and R is the generic k-algebra, cf. [25, 56].

We next study the validity of Axiom 3. As in I §§19 and 20, we shall call an object J in a cartesian closed category $\mathcal{E}$ an atom if $(-)^J: \mathcal{E} \to \mathcal{E}$ has a right adjoint.

In functor categories $\underline{Sets}^{\mathcal{R}}$ ($\mathcal{R}$ any small category), there are many atoms:

Proposition 1.4. If $B \in \mathcal{R}$ has the property that a coproduct $B+B'$ exists in $\mathcal{R}$ for any $B' \in \mathcal{R}$, then the representable functor $y(B) \in \underline{Set}^{\mathcal{R}}$ is an atom.

Proof. For any $F \in \underline{Set}^{\mathcal{R}}$,

$$(F^{yB})(B') = \hom(yB', F^{yB}) = \hom(yB \times yB', F) = \hom(y(B+B'), F) \tag{1.5}$$

(since the Yoneda embedding $\mathcal{R}^{op} \to \underline{Set}^{\mathcal{R}}$ preserves those limits that exist)

$$= F(B + B')$$

so that

$$F^{yB} = F \circ (B + -).$$

It is well known (cf. e.g.[55] X.3, Coroll.2) that functors between categories of form $\underline{\text{Set}}^{\mathbb{C}}$, which are induced by functors between the index categories, have adjoints on both sides. But the above calculation shows that $(-)^{yB}$ is induced by $B + -: \mathcal{R} \longrightarrow \mathcal{R}$.

Proposition 1.5. If $\mathcal{R}$ and R are as in Theorem 1.2, and $\mathcal{E} = \underline{\text{Set}}^{\mathcal{R}}$, then, for any $B \in FG\mathbb{T}_k$ which is stable with respect to $\mathcal{R}$, $\text{Spec}_R(B)$ is an atom.

Proof. Even though B is not an object of $\mathcal{R}$, we still have the functor $B \otimes_k - : \mathcal{R} \to \mathcal{R}$, by the stability. The proof is now almost as the proof of the preceding Proposition; now we have to see that

$$F^{\text{Spec}_R(B)} = F \circ (B \otimes_k -),$$

and this calculation is as (1.5) above, except we have to utilize Lemma 1.1.

Proposition 1.4 has the following Corollary:

Theorem 1.6. The model considered in Corollary 1.3 satisfies Axiom 3_k.

Whereas it was easy to transfer validity of Axioms 1 and 2 from $\underline{\text{Set}}^{\mathcal{R}}$ to suitable subcategories, it is usually more subtle and requires more assumptions to produce atoms in such subcategories, in particular to prove Axiom 3_k. We refer the reader to [42] to see how it sometimes may be done; cf. also the end of §8.

We finish this § with some remarks that are intended to clarify how the notion of 'elements defined at different stages' (cf. Part II) in categories of form $\underline{\text{Set}}^{\mathcal{R}}$ (with $\mathcal{R}$ a category of rings) are related to 'elements of geometric objects, elements which are defined at various rings, or with varying degree of reality'. We do it with an example.

Let $B = \mathbb{Z}[X,Y]/(X^2+Y^2-1)$. For any ring object R in a category $\mathcal{E}$ with finite inverse limits,

$$\mathrm{Spec}_R(B) = [\![(x,y) \in R \times R \mid x^2 + y^2 = 1]\!] ,$$

"the circle over R"; denote it $S^1(R)$. For the special case where $\mathcal{E} = \underline{\mathrm{Set}}^{\mathcal{R}}$, where $\mathcal{R}$ is any suitable category of rings, an element a of $S^1(R)$, defined at stage yC,

$$a \in_{yC} S^1(R),$$

is the same thing, by Yoneda's Lemma, as an element of $\hom_{\mathrm{Rings}}(B,C)$, i.e. an element of the set

$$\{(x,y) \in C \times C \mid x^2 + y^2 = 1\} = S^1(C) ,$$

the "set theoretical circle over C". The standard unit circle, which comes about by taking $C = \mathbb{R}$, thus consists of certain elements of $S^1(R) \subseteq R \times R$ defined at stage $y\mathbb{R}$, the '<u>real</u> points of the circle' in classical terminology; this terminology also allows for '<u>complex</u> points' (or '<u>imaginary</u> points') of the circle, e.g. its two points at infinity, or to give an example not involving projective geometry, the two 'imaginary' points common to the unit circle and the line $x = 2$, say. In our context, they are simply elements of $S^1(R)$ defined at stage $y\,\mathbb{C}$ (which is a stage 'later' than $y\,\mathbb{R}$, due to $y(i): y\,\mathbb{C} \to y\,\mathbb{R}$ induced by the standard embedding $i: \mathbb{R} \hookrightarrow \mathbb{C}$).

<u>EXERCISES</u>

1.1. Let $\mathcal{R}$ be any full subcategory of the category of k-algebras, and let $\mathcal{E} = \mathrm{Set}^{\mathcal{R}}$, R = forgetful functor. For $B \in \mathrm{FG}\mathbb{T}_k$, interpret B as a ring of functions $\mathrm{Spec}_R(B) \longrightarrow R$. (Hint: $b \in B$ gives rise to $F(1) \to B$ (where $F(1) = k[X]$); apply Spec_R to it and use $\mathrm{Spec}_R(F(1)) = R$.) Use this to construct for each $B' \in \mathcal{R}$, a map $j: B \to R^{\mathrm{Spec}_R(B)}(B')$.

1.2. With $\mathcal{E}$, R, and B as in Exercise 1, and $C \in$ R-Alg, describe a map

$$\underset{\sim}{\mathrm{Hom}}_{R\text{-Alg}}(R^{\mathrm{Spec}_R(B)}, C) \xrightarrow{\mu_{B,C}} \mathrm{Spec}_C(B)$$

as follows. Give its B'-component ($B' \in \mathcal{R}$) by associating to the data (1.3) the map

$$B = \mathbf{1} \times B \xrightarrow{id_{B'} \times j} y(B')(B') \times R^{\mathrm{Spec}_R(B)}(B') \xrightarrow{\tau_{B'}} C(B')$$

with j as in Exercise 1. Interpret this map as an element of $\mathrm{Spec}_C(B)(B')$ (as in the proof of Theorem 1.2).

1.3. Prove that the $\mu_{B,C}$ constructed in Exercise 2 is natural in $B \in FG\Pi_k$, and, for $B = F(n)$, agrees with the $\nu_{B,C}$ of I.(12.2). Conclude by Appendix A that $\mu_{B,C} = \nu_{B,C}$.

1.4. If B is stable with respect to $\mathcal{R}$, the $R^{\mathrm{Spec}_R(B)}(B')$ may be identified with $B' \otimes_k B \in \mathcal{R}$ (cf. the proof of Theorem 1.2), and under this identification, the map j of Exercise 1 becomes the inclusion $B \to B' \otimes_k B$. Conclude that the process described in the proof of Theorem 1.1 to lead from (1.2) to (1.4) is just $\nu_{B,C}$.

III.2: MODELS FOR ε-STABLE GEOMETRIC THEORIES

We shall here prove that certain of the models $\mathcal{E}$, R described in the previous § satisfy not only Axiom 1^W and Axiom 3, but also some first order sentences like

$$\forall x: (x \text{ is invertible}) \vee ((1-x) \text{ is invertible})$$

(which, together with $\neg(0 = 1)$, express: R is a local ring). To be able to interpret such sentences which involve $\exists, \vee,$ and $\neg$, we must have a Grothendieck topology (cf. II. §8) on $\mathcal{E}$.

The models here will be of form $\mathcal{E} = \underline{\text{Set}}^{\mathcal{R}}$, with $\mathcal{R}$ some small full subcategory of the category of k-algebras, and R: $\mathcal{R} \to$ <u>Set</u> will be the forgetful functor. The Grothendieck topology we consider on $\mathcal{E}$ is the canonical one (cf. II.§9). It can be described by the pretopology given by

$$\{F_i \to F \mid i \in I\} \in \text{Cov}(F)$$

if for each $C \in \mathcal{R}$, the family of set-theoretic mappings

$$\{F_i(C) \to F(C) \mid i \in I\}$$

is jointly surjective. The representable functors, y(B) for $B \in \mathcal{R}$, form a dense set $\mathcal{A}$ of generators, which is topologically dense. We know from II.§8 that we only have to consider $\mathcal{A}$-elements to describe satisfaction $\vdash$.

For a family

$$\{y(f_i): y(B_i) \to y(B) \mid i \in I\}$$

to be a covering, a necessary and sufficient condition is that for some $i \in I$, $y(f_i)$ is split epic, or equivalently, that $f_i: B \to B_i$ is split mono, i.e. has a left inverse g. (For the necessity, apply the 'jointly surjective' criterion for $C = B$.) None of the representables are covered by an empty family.

One reason why geometric formulae and theories are important in our context is because of the following two Propositions. Note first that, for any $B \in \mathcal{R}$, we have, by Yoneda's lemma, a bijective correspondence between maps $b: yB \to R$, and elements $\bar{b} \in R(B) = B$ (i.e. to $b \in_{yB} R$ corresponds $\bar{b} \in B$).

Proposition 2.1. Suppose $\varphi(x_1,\ldots,x_n)$ is a geometric formula. For any n-tuple

$$yB \xrightarrow{b_i} R \qquad (i = 1,\ldots,n),$$

we have

$$\vdash_{yB} \quad \varphi(b_1,\ldots,b_n)$$

if and only if

$$\varphi(\bar{b}_1,\ldots,\bar{b}_n) \text{ holds in } B.$$

Proof. This is proved by induction in the way φ is built from polynomial equations by means of $\wedge$, $\vee$, and $\exists$. For the polynomial equations, it follows immediately, because the ring structure on R was described argumentwise, so that $\hom(yB,R) \cong R(B)$ is a ring isomorphism. Let us do the induction step for $\exists$, assuming for simplicity that $n = 1$, and $\varphi(b)$ is $\exists z\ \psi(b,z)$, where the $\psi(x,y)$ is a geometric formula to which the induction hypothesis applies. So assume $\vdash_{yB} \varphi(b)$, i.e.

$$\vdash_{yB} \quad \exists z: \psi(b,z) .$$

We can thus find a covering $\{yB_i \longrightarrow yB \mid i \in I\}$ of yB, and elements $c_i\colon yB_i \to R$ with $\vdash_{yB_i} \psi(b,c_i)$ for all i. One of the $yB_i \to yB$ is of form $y(f)\colon y(C) \longrightarrow y(B)$ with $f\colon B \to C$ a split monic, $g \circ f = id_B$, and we have $c\colon yC \to R$ with $\vdash_{yC}\psi(b,c)$. Changing stage along $y(g)\colon y(B) \to y(C)$, we thus also have $\vdash_{yB} \psi(b,c)$, (note that we get our original b back since $g \circ f = id_B$). By the induction hypothesis, we thus have $\psi(\bar{b},\bar{c})$. Hence also $\exists z\ \psi(\bar{b},z)$ holds in B. The other implication is trivial. Disjunction is treated similarly; conjunction is trivial.

Proposition 2.2. Let $\mathcal{R}$ be a small full subcategory of the category of $\mathbb{T}_k$-algebras, and assume all $B \in \mathcal{R}$ satisfy the axioms of a geometric theory $\mathcal{T}$. Then $R \in \underline{Set}^{\mathcal{R}}$ (= the forgetful functor) is a model of $\mathcal{T}$.

Proof. Let

$$\forall x_1,\dots,x_n\colon \varphi(x_1,\dots,x_n) \Rightarrow \psi(x_1,\dots,x_n)$$

be one of the Axioms of $\mathcal{T}$. To prove that

$$\vdash_1 \forall x_1,\dots,x_n\colon \varphi(x_1,\dots,x_n) \Rightarrow \psi(x_1,\dots,x_n) \tag{2.1}$$

holds for R, let

$$yB \xrightarrow{b_i} R \qquad i = 1,\dots,n$$

be an n-tuple of A-elements of R with

$$\vdash_{yB} \varphi(b_1,\dots,b_n).$$

Since φ is a geometric formula, Proposition 2.1 yields that $\varphi(\bar{b}_1,\dots,\bar{b}_n)$ holds in B. Since B satisfies $\mathcal{T}$, we therefore also have $\psi(\bar{b}_1,\dots,\bar{b}_n)$ holding in B, so by Proposition 2.1 again

$\vdash_{yB} \psi(b_1,\ldots,b_n)$.

This proves (2.1).

Theorem 2.3. If $\mathcal{R}$ is the category of those (finitely generated, say) k-algebras which are local rings, then, for $\mathcal{E} = \underline{\mathrm{Set}}^{\mathcal{R}}$ and R = the forgetful functor $\mathcal{R} \to \underline{\mathrm{Set}}$, we have that R is a local ring object, and Axiom 1_k^W and Axiom 3_k hold. Also, for each $n = 1,2,\ldots$

$$\vdash_1 \forall x_1,\ldots,x_n: \quad \neg(\bigwedge_{i=1}^{n} (x_i = 0)) \Rightarrow (\bigvee_{i=1}^{n} (x_i \text{ is invertible})). \tag{2.2}$$

Proof. The last assertion is an "extra" which falls outside the general theory, and we refer to Exercise 2.2 for a proof. The other assertions are part of a general principle, whose pillars are §1 and Proposition 2.2. The latter immediately gives that R is a local ring object. To prove Axioms 1_k^W and 3_k, using §1, we need

Lemma 2.4. If B is a k-algebra which is a local ring, and W is a Weil algebra over k, then $B \otimes_k W$ is a local ring. In particular, W is stable with respect to $\mathcal{R}$.

Proof. Let $W = k \oplus k^{n-1} = k \oplus W'$ with all elements in W' nilpotent. Then $B \otimes_k W = B \oplus B'$ with all elements in B' nilpotent. Thus, an element $(b,b') \in B \oplus B'$ is invertible iff $b \in B$ is invertible. To say that B is local can be expresssed: the non-invertibles are stable under addition. Then it is clear that the non-invertibles in $B \oplus B'$ are stable under addition.

Having the Lemma, we have that $\mathrm{Spec}_R(W)$ is an atom for any Weil algebra W, by Proposition 1.5, so Axiom 3_k holds. From Theorem 1.2, we conclude that for any Weil algebra W and for any R-algebra C in $\mathcal{E}$, the map $\nu_{W,C}$

$$\underset{\sim\sim}{\mathrm{Hom}}_{R\text{-}Alg}(R^{\mathrm{Spec}_R W}, C) \longrightarrow \mathrm{Spec}_C(W)$$

is an isomorphism, from which Axiom 1_k^W follows, cf. the proof of Theorem I.16.1. This proves the theorem.

Let us agree to call a geometric theory $\mathcal{T}$ in the language of the theory of k-algebras <u>ε-stable</u> if it has the property: if B is a model of $\mathcal{T}$, then so is $B \otimes_k W$, for any Weil algebra W over k. The theory of local rings (or local k-algebras) is ε-stable, by Lemma 2.4. We immediately see that the proof given for Theorem 2.3 also can be used to prove

<u>Theorem 2.5</u>. Let $\mathcal{T}$ be an ε-stable geometric theory of k-algebras. Then there exists a model $(\mathcal{E},R)$ of Axiom 1_k^W, Axiom 3_k, and $\mathcal{T}$.

We take in fact $\mathcal{E} = \underline{Set}^{\mathcal{R}}$ for $\mathcal{R}$ some small category of $\mathcal{T}$-models such that $\mathcal{R}$ is stable under functors of form $-\otimes_k W$ for any Weil algebra W over k.

A more subtle theorem [6], [14] is that if $\mathcal{T}$ is ε-stable, then the <u>generic</u> $\mathcal{T}$-model (in the sense of classifying toposes, cf. [56]) satisfies Axioms 1_k^W, 3_k, (and $\mathcal{T}$ of course). These models live in toposes of sheaves on $(FP\mathbb{T}_k)^{op}$, for a Grothendieck topology $j_{\mathcal{T}}$ on this category, which is defined in terms of the axioms of the theory $\mathcal{T}$. It is easy to prove that if $j_{\mathcal{T}}$ is subcanonical, then the argument of Theorem 1.2 and Corollary 1.3 yields also that Axiom 2_k holds for R in $\mathcal{E}$ (R being the forgetful functor $FP\mathbb{T}_k \to \underline{Set}$). We collect some of the information given here:

<u>Theorem 2.6</u>. Let $\mathcal{T}$ be an ε-stable geometric theory of k-algebras. Then Axioms 1_k^W and 3_k hold for R = generic $\mathcal{T}$-model. If the Grothendieck topology $j_{\mathcal{T}}$ is subcanonical, Axiom 2_k holds as well.

The theorem (including the last part) applies in particular to the theory of k-algebras which are local rings, and to the theory of separably closed local k-algebras (cf. Exercise 2.6 below for more information about this notion). The first part of the Theorem applies to the theory $\mathcal{T}_r$ of Exercise 2.6.

EXERCISES

2.1. Let $\mathcal{R}$ be any small category of rings, and $\mathcal{E} = \underline{\text{Set}}^{\mathcal{R}}$, R = forgetful functor. Let $y: \mathcal{R}^{op} \to \underline{\text{Set}}^{\mathcal{R}}$ be the Yoneda embedding. Prove, for $b_1,\ldots,b_n \in_{y(B)} R$ that

$$\vdash_{y(B)} \quad \neg(b_1 = 0 \wedge \ldots \wedge b_n = 0)$$

if and only if: for no $f: B \to C$ in $\mathcal{R}$ do we have $f(\bar{b}_1) = \ldots = f(\bar{b}_n) = 0$ (notation as in Proposition 2.1).

2.2. If B is a local ring, prove that if none of $b_1,\ldots,b_n \in B$ are invertible, then $B/(b_1,\ldots,b_n)$ is a local ring. Combine this with Exercise 1 to conclude that for the model $(\mathcal{E},R)$ considered in Theorem 2.3, we have, for each n, validity of (2.2).

2.3. Prove that the geometric field notion II.(9.7) is not ε-stable.

2.4. Prove that the theory of k-algebras satisfying II.(9.8) is ε-stable, provided k contains the rational numbers. (Hint: Write $B \otimes_k W$ as $B \oplus B'$, as in the proof of Lemma 2.4. To find $\sqrt{(b,b')}$ one may use Taylor series for $\sqrt{\ }$, developed from $\sqrt{(b,0)}$.

2.5. Prove that the theory of k-algebras satisfying

$$\forall x: (\exists y: y^2 = x \vee \exists y: y^2 = -x)$$

is not ε-stable.

2.6. Consider the theory $\mathcal{T}$ of rings satisfying

$$\forall x_1,\ldots,x_n: (\bigvee_{i=1}^{n} (x_i \text{ is invertible})) \Rightarrow (\sum x_i^2 \text{ is invertible})$$

for all $n = 1,2,\ldots$. Prove that $\mathcal{T}$ is ε-stable, and that, if B is a local ring which satisfies $\mathcal{T}$, then

$$B[i] := B[X]/(X^2+1)$$

is a local ring. (Models of $\mathcal{T}$ are the so-called "formally-real" rings.)

There exists a geometric theory $\bar{\mathcal{T}}$ (Joyal and Wraith [81]), whose models in Sets are the Henselian local rings with separably closed residue field. There is also [33] a geometric theory $\mathcal{T}_r$ whose models are those formally real local rings such that B[i] is a model of $\bar{\mathcal{T}}$. The models in Sets of $\mathcal{T}_r$ are Henselian local rings with real-closed residue field. Models of $\bar{\mathcal{T}}$, respectively $\mathcal{T}_r$, are the so-called separably closed local rings, and separably real-closed local rings. These notions are ε-stable substitutes for the (not ε-stable) notions of algebraically closed, respectively real closed field.

2.7. Formulate by geometric sentences the axioms for the theory of formally real local rings in which each invertible square sum has a square root ("Pythagorean local rings"), and prove this theory ε-stable.

2.8. For theories formulated in the language of the theory of k-algebras and a preorder relation $\leq$, we define the notion of ε-stability as above, and by further declaring the order relation $\leq$ on $B \otimes_k W$ in terms of the order relation $\leq$ on B by

$$(b_1,b_1') \leq (b_2,b_2') \text{ iff } b_1 \leq b_2$$

(identifying $B \otimes_k W$ with $B \oplus B'$ as in the proof of Lemma 2.4).

Prove that the theory of preordered rings considered in I §13 is ε-stable, but that ε-stability is lost if we further require $\leq$ to be antisymmetric.

2.9. Prove that the theory of Archimedean preordered rings (meaning the theory described by II.(9.9)) is ε-stable.

III.3: AXIOMATIC THEORY OF WELL-ADAPTED MODELS (1)

For the first time in these notes, we shall now presuppose classical differential calculus, and manifold theory; in particular, we shall consider the real numbers, $\mathbb{R}$. We do it in order to answer the question what a comparison between classical analysis and the synthetic theory developed here should be like, to justify the borrowing of names ("derivative", "vector field", "Lie-bracket", etc.).

In the following, the word manifold means smooth (= C^∞) manifold, Hausdorff and with a countable basis (and hence paracompact and with partitions of unity, cf. §5 below). Let Mf denote the category of these (with C^∞ maps as morphisms). It is a small category.

The category Mf has rather poor category theoretic properties, in particular, it does not have all finite inverse limits, and some of those finite inverse limits it does happen to have, are "wrong"; for example, in Mf, the intersection between the x-axis in $\mathbb{R}^2$ and a tangent unit circle exists in Mf, and is the (unique) manifold with just one point, whereas, from our point of view (and Protagoras', Hjelmslev's, Grothendieck's) this intersection should be a certain bigger object D. Those intersections in Mf, which have good properties from all viewpoints, are the transversal ones. More generally, Mf has good transversal pullbacks; we recall

Definition 3.1. A pair of maps $f_i: M_i \to N$ $(i = 1,2)$ in Mf with common codomain are said to be transversal to each other if for each pair of points $x_1 \in M_1$, $x_2 \in M_2$ with $f_1(x_1) = f_2(x_2)$ (= y, say), the images of $(df_i)_{x_i}$ $(i = 1,2)$ jointly span T_yN as a vector space.

This is easily seen to be equivalent to saying that the map

$$f_1 \times f_2 : M_1 \times M_2 \longrightarrow N \times N$$

is transversal to the submanifold $\Delta: N \hookrightarrow N \times N$. From this, and the "preimage-theorem" (see e.g. [24], p.21) follows that $(f_1 \times f_2)^{-1}(\Delta)$ is a submanifold of $M_1 \times M_2$. But it sits in a set-theoretic pull-back diagram

$$\begin{array}{ccc} (f_1 \times f_2)^{-1}(\Delta) & \longrightarrow & M_1 \\ \downarrow & & \downarrow f_1 \\ M_2 & \xrightarrow[f_2]{} & N \end{array}$$

which is therefore also a pull-back diagram in Mf. Such pull-back diagrams we call <u>transversal</u>.

When we say that a functor from Mf into a category E preserves transversal pull-backs, we mean that any transversal pull-back in Mf is transformed into a pull-back diagram in E, and that the terminal object **1** in Mf (= the one-point manifold) goes to the terminal object in E.

Given $f: M \to N$ in Mf. To say that $y \in N$ is a <u>regular value</u> (standard terminology, cf. e.g.[24],p.21) may be expressed: the maps $\ulcorner y \urcorner : \mathbf{1} \to N$ and $f: M \to N$ are transversal. This is in particular the case if y is not in the image of f. If all $y \in N$ are regular values, one says that f is a <u>submersion</u>. In this case, any smooth map with codomain N is transversal to f. An open inclusion is evidently a submersion.

Also, a product of two manifolds M_1 and M_2 may be viewed as a transversal pull-back of M_1 and M_2 over **1**.

In the following, we shall consider a cartesian closed category E with finite inverse limits, and a functor

$$i: Mf \longrightarrow E .$$

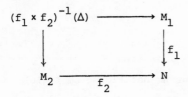

The data for a well-adapted model is just such $\mathcal{E}$ and i. These data are required to satisfy Axioms A and B, to be described now, and Axioms C and D to be described in the next §.

Axiom A. The functor i preserves transversal pull-backs.

Since $\mathbb{R}$ is an $\mathbb{R}$-algebra object in Mf, and i preserves finite products, we get from Axiom A that $i\,\mathbb{R}$ is an $\mathbb{R}$-algebra object in $\mathcal{E}$. This object, we denote R:

$$R := i(\mathbb{R}) .$$

It now makes sense to state (using notation from I §16):

Axiom B (= Axiom $1^W_{\mathbb{R}}$). For any Weil algebra W over $\mathbb{R}$, the R-algebra homomorphism

$$\alpha\colon R \otimes W \longrightarrow R^{\mathrm{Spec}_R(W)}$$

is an isomorphism.

In the rest of this §, we assume Axioms A and B. An Axiom B.2 stronger than Axiom B is presented in some of the exercises of §9.

Since Axiom $1^W_{\mathbb{R}}$ implies Axiom 1 (compare I §12 and I §16), it follows that for any given map $g\colon R \to R$ in the category $\mathcal{E}$, we can define its ("synthetic") derivative $g'\colon R \to R$, as in I §2. We have the following comparison theorem between synthetic differentiation and the usual "analytic" one:

Theorem 3.2. Let $f\colon \mathbb{R} \longrightarrow \mathbb{R}$ be a smooth map. Then

$$(i(f))' = i(f').$$

Proof. A well known theorem from analysis ("Hadamard's lemma") says that if $f: \mathbb{R} \longrightarrow \mathbb{R}$ is smooth, then the function $f(x+t)$ in the two variables x and t can be written

$$f(x+t) = f(x) + t \cdot f'(x) + t^2 \cdot g(t,x), \tag{3.1}$$

where $g: \mathbb{R} \times \mathbb{R} \to \mathbb{R}$ is some uniquely determined smooth function. Now the validity of (3.1) can be expressed in terms of commutativity of a certain diagram starting in $\mathbb{R} \times \mathbb{R}$, built by means of cartesian products, from f, f', and g, and the ring operations 'plus' and 'times'. The functor $i: Mf \longrightarrow E$ preserves such products (by Axiom A), and takes the ring operations of $\mathbb{R}$ into those of $i(\mathbb{R}) = R$, by construction of the ring structure of the latter. Thus we get a commutative diagram in E, starting in $R \times R$. The commutativity of it can, by the extensionality principle for maps (Proposition II.3.2) be expressed

$$\vdash_1 \forall x,t: (i(f))(x+t) = (i(f))(x) + t \cdot (i(f'))(x) + t^2 \cdot (i(g))(t,x)$$

From this, we deduce

$$\vdash_1 \forall x \in R \ \forall d \in D: \ (i(f))(x+d) = (i(f))(x) + d \cdot (i(f'))(x). \tag{3.2}$$

Let $x \in_X R$ be a generalized element of R and $d \in_Y R$ a generalized element of D, defined at the later stage $\alpha: Y \to X$. From (3.2), we get

$$\vdash_Y (i(f))(x+d) = (i(f))(x) + d \cdot (i(f'))(x),$$

but by Taylor's formula, Theorem I 2.1, applied for $i(f)$ and the given (generalized) elements x and d, we have

$$\vdash_Y (i(f))(x+d) = (i(f))(x) + d \cdot (i(f))'(x).$$

Subtracting these two equations (which hold in the ring $hom(Y,R)$), we get

$$\vdash_Y (d\cdot(i(f'))(x) = d\cdot(i(f))'(x).$$

Since α, d were arbitrary, we get

$$\vdash_X \forall d \in D\colon d\cdot(i(f'))(x) = d\cdot(i(f))'(x),$$

and cancelling the universally quantified d, (cf. Exercise II.5.2) we get

$$\vdash_X (i(f'))(x) = (i(f))'(x).$$

Since x, X were arbitrary, we conclude from the extensionality principle for maps that $i(f') = (i(f))'$.

Of course, even when $i\colon Mf \to \mathcal{E}$ is full, this theorem talks about actual maps $R \to R$ in $\mathcal{E}$, or equivalently, global sections of R^R. There is a more general form talking about elements $g \in_{i(M)} R^R$,

$$i(M) \xrightarrow{\;g\;} R^R \quad ;$$

namely, if the exponential adjoint of it,

$$i(M \times \mathbb{R}) = i(M) \times R \longrightarrow R = i(\mathbb{R})$$

comes about as $i(f)$, for $f\colon M \times \mathbb{R} \to \mathbb{R}$ a smooth map, then one may similarly prove (cf. Exercise 3.2) that

$$g' = \left(i\left(\frac{\partial f}{\partial t}\right)\right)^{\wedge}, \tag{3.3}$$

where $\frac{\partial f}{\partial t}$ means partial differentiation after the last variable, and $\wedge$ denotes exponential adjoint.

The technique used for proving Theorem 3.2 may be used to prove also

<u>Theorem 3.3</u>. Let $g\colon \mathbb{R}^n \to \mathbb{R}$ be smooth. Then:

$$i\left(\frac{\partial g}{\partial x_j}\right) = \frac{\partial(i(g))}{\partial x_j} \qquad \text{for } j = 1,\dots,n,$$

and hence also: i commutes with formation of Jacobian

$$i(dg) = d(i(g))\colon R^n \times R^n \longrightarrow R^n .$$

(Recall that $dg(\underline{x},\underline{y}) = dg_{\underline{x}}(\underline{y}) = \sum \frac{\partial g}{\partial x_i}(\underline{x})\cdot y_i$, and similarly in the synthetic setting).

This theorem also holds in parametrized form, in analogy with (3.3).

Recall from I §19 that if B is any ring object in a category with sufficiently many finite inverse limits, then $\mathrm{Inv}(B) \hookrightarrow B$ is the subobject defined as the upper composite in the diagram

$$\begin{array}{ccccc} \mathrm{Inv}(B) & \longrightarrow & B \times B & \xrightarrow{\ \mathrm{proj}_1\ } & B \\ \downarrow & * & \downarrow{\scriptstyle m} & & \\ \mathbf{1} & \xrightarrow[\ulcorner 1 \urcorner]{} & B & & \end{array}$$

where m is the multiplication, and the square $*$ is formed as a pull-back.

If $B = \mathbb{R}$ in the category Mf, the pull-back diagram $*$ exists and is in fact transversal: if x and $y \in \mathbb{R}$ have $x\cdot y = 1$, then

$$dm_{(x,y)} = \{y,x\} ,$$

which is a matrix of rank one, since x is invertible. (Equivalently, 1 is a regular value for the multiplication map.) It follows that $i\colon Mf \to \mathcal{E}$ preserves this pull-back, so that we get, in $\mathcal{E}$,

$$i(\mathrm{Inv}(\mathbb{R})) = \mathrm{Inv}(R),$$

as subobjects of R. (Of course, $\mathrm{Inv}(\mathbb{R}) = \{x \in \mathbb{R} \mid x \neq 0\}$.) Also, Axiom $1^W_{\mathbb{R}}$ implies that $\mathrm{Inv}(R) \hookrightarrow R$ is formal-étale in the sense of I §17, by Proposition I.19.1.

We can now prove

Theorem 3.4. If U is an open subset of a manifold M, the inclusion $U \hookrightarrow M$ goes by the functor i to a formal étale monic $iU \longrightarrow iM$ in $\mathcal{E}$.

Proof. A well-known theorem of analysis (see e.g. [24] Ex. 1.5.11, and combine with a partition-of-unity argument) says that any open subset $U \subseteq M$ of a manifold is of form $f^{-1}(\mathrm{Inv}(\mathbb{R}))$ for some smooth $f\colon M \to \mathbb{R}$. Such f we may call a smooth characteristic function for U. The pull-back diagram defining $f^{-1}(\mathrm{Inv}(\mathbb{R}))$ is transversal because $\mathrm{Inv}(\mathbb{R}) \hookrightarrow \mathbb{R}$ is an open inclusion, hence a submersion. Since i preserves such pull-backs, we get a pull-back square in $\mathcal{E}$

$$\begin{array}{ccc}
iU & \rightarrowtail & iM \\
\downarrow & & \downarrow \\
\mathrm{Inv}(R) = i(\mathrm{Inv}(\mathbb{R})) & \rightarrowtail & i\mathbb{R} = R\,.
\end{array}$$

But the class of formal-étale maps is stable under pull-back, by Proposition I.19.2.

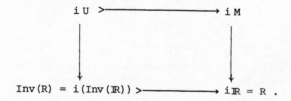

Recalling the terminology of I §19, we see that the proof also gives

<u>Proposition 3.5</u>. If U is an open subset of a manifold M, $\mathrm{i}\,U \rightarrowtail \mathrm{i}\,M$ is strongly étale (or, is an open inclusion). Similarly for injective maps which map onto an open subset of M ("open embeddings in Mf").

<u>EXERCISES</u>

3.1. Prove that if the two squares in a diagram

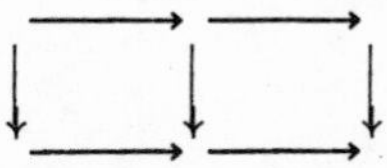

in Mf are transversal pull-backs, then so is the total diagram.

3.2. For $M \in \mathrm{Mf}$, a smooth map $f: M \times \mathbb{R} \to \mathbb{R}$ may be viewed as a smoothly parametrized family of maps $\{f_m: \mathbb{R} \to \mathbb{R} \mid m \in M\}$. Hadamard's lemma is known to be "smooth in parameters", meaning that if we for each $m \in M$ take the $g_m: \mathbb{R}^2 \to \mathbb{R}$ making (3.1) true, then $g: M \times \mathbb{R}^2 \longrightarrow \mathbb{R}$ given by $(m,t,x) \longmapsto g_m(t,x)$, is smooth. Use this to prove (3.3).

III.4: AXIOMATIC THEORY OF WELL-ADAPTED MODELS (2)

Unlike Axioms A and B in the preceding §, the Axioms C and D to be presented now deal with colimit- and covering-related notions. To state Axiom C, we need the concepts of <u>pretopology</u> and (Grothendieck-) <u>topos</u> (cf. Appendix B).

The category Mf has a subcanonical pretopology whose coverings

$$\{M_j \longrightarrow M \mid j \in J\}$$

are jointly surjective families of open inclusions. Such families we call, of course, <u>open coverings</u>.

<u>Axiom C</u>. $\mathcal{E}$ is a topos, and the functor $i: \mathrm{Mf} \to \mathcal{E}$ takes open coverings to coverings (briefly "i preserves coverings").

Recall that a <u>covering</u> in a topos $\mathcal{E}$. is simply a jointly epic family; these define the canonical topology on it, which is in fact subcanonical.

The most important consequence of Axioms A, B, C is the following theorem, which one may rightfully demand of any comparison between the synthetic and the analytic theory. Assuming these three axioms, we have

<u>Theorem 4.1</u>. The functor $i: \mathrm{Mf} \to \mathcal{E}$ commutes with formation of tangent-bundle, i.e. there exists an isomorphism, natural in $M \in \mathrm{Mf}$,

$$i(TM) \xrightarrow[\cong]{\alpha_M} T(iM) = (iM)^D,$$

where TM is the classical tangent bundle; and for $M = \mathbb{R}$, this α is the α of I.(1.1)

$$i(T\,\mathbb{R}) = i(\mathbb{R}\times\mathbb{R}) = R\times R \xrightarrow{\alpha} R^D \tag{4.1}$$

There is a more general version of the Theorem, involving the "Weil prolongation of type W" [79] for any Weil-algebra W over $\mathbb{R}$, and which states the existence of a natural isomorphism

$$i(T_W M) \xrightarrow{\alpha_M} (iM)^{\mathrm{Spec}_R(W)}$$

where $T_W M$ is the Weil-prolongation of type W (in Weil's notation $^W M$), and with $\alpha_{\mathbb{R}}$ equal to the α of Axiom $1^W_{\mathbb{R}}$ (I §16). We refer the reader to [36] for this generalization. The case stated in the Theorem corresponds to the case $W = \mathbb{R}[\varepsilon]$.
- The functors $T_W\colon \mathrm{Mf} \to \mathrm{Mf}$ (in particular, the tangent bundle functor) takes transversal pull-backs to transversal pull-backs, and open coverings to open coverings.

Proof of Theorem 4.1. Since both T, i, and $(-)^D$ commute with products, we get from (4.1) and Axiom B an isomorphism $\alpha_{\mathbb{R}^n}\colon i(T(\mathbb{R}^n)) \longrightarrow (R^n)^D$. If $U \hookrightarrow \mathbb{R}^n$ is an open subset, we have a transversal pull-back in Mf,

$$\begin{array}{ccc} TU & \hookrightarrow & T\,\mathbb{R}^n \\ \downarrow & & \downarrow \\ U & \hookrightarrow & \mathbb{R}^n \end{array} ,$$

and hence the first of the two diagrams

$$\begin{array}{ccc} i(TU) & \longrightarrow & i(T\,\mathbb{R}^n) \\ \downarrow & & \downarrow \\ iU & \longrightarrow & R^n \end{array} \qquad \begin{array}{ccc} (iU)^D & \longrightarrow & (R^n)^D \\ \downarrow & & \downarrow \\ iU & \longrightarrow & R^n \end{array}$$

is a pull-back. The second diagram is a pull-back because $iU \to R^n$ is formal-étale, by Theorem 3.4. Since the upper right hand corners are isomorphic via $\alpha_{\mathbb{R}^n}$, the two upper left hand corners are isomorphic, by a map

$$i\,T\,U \xrightarrow{\alpha_U} (i\,U)^D \tag{4.2}$$

which is at least natural with respect to the inclusion map $U \subseteq \mathbb{R}^n$.

The strategy of the proof is now 1) to prove the α_U's natural with respect to any smooth map between open subsets of the $\mathbb{R}^n$'s (this is the core of the Theorem), and then 2), to use a patching argument to get a natural α_M, for all $M \in \mathrm{Mf}$. Finally 3), we prove this α_M invertible for all $M \in \mathrm{Mf}$.

We first generalize Proposition I.1.1, by reformulating it in terms of algebraic theories (cf. Appendix A, and Exercise I.1.4). We consider the algebraic theory $\mathbb{T}_\infty$ whose n-ary operations are the smooth ($= C^\infty$) maps $\mathbb{R}^n \longrightarrow \mathbb{R}$. Since polynomials with real coefficients are among these, any $\mathbb{T}_\infty$-algebra is also an $\mathbb{R}$-algebra. Evidently, $\mathbb{R}$ is an algebra for $\mathbb{T}_\infty$, but also $\mathbb{R} \times \mathbb{R} = \mathbb{R}[\varepsilon]$ carries a canonical structure of algebra for $\mathbb{T}_\infty$, extending its already existing structure of $\mathbb{R}$-algebra. Namely, for $g: \mathbb{R}^n \to \mathbb{R}$ an n-ary operation of $\mathbb{T}_\infty$, we define its action $g_{\mathbb{R}[\varepsilon]}$: $(\mathbb{R}[\varepsilon])^n \longrightarrow \mathbb{R}[\varepsilon]$ on $\mathbb{R}[\varepsilon]$ by

$$(x_1+\varepsilon\cdot y_1,\dots,x_n+\varepsilon\cdot y_n) \longmapsto g(x_1,\dots,x_n) + \varepsilon\cdot\sum \frac{\partial g}{\partial x_i}(x_1,\dots,x_n)\cdot y_i \,,$$

or equivalently, identifying $(\mathbb{R}[\varepsilon])^n = (\mathbb{R}^2)^n$ with $\mathbb{R}^n \times \mathbb{R}^n$,

$$(\underline{x},\underline{y}) \longmapsto (g(\underline{x}), dg_{\underline{x}}(\underline{y})). \tag{4.3}$$

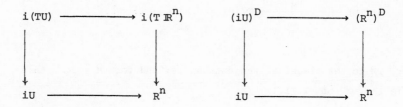

Since the functor $i\colon Mf \to \mathcal{E}$ preserves products, it follows that $i(\mathbb{R}) = R$, as well as $i(\mathbb{R}[\varepsilon]) = R[\varepsilon]$, are algebras for $\mathbb{T}_\infty$, and since $(-)^D$ preserves products, R^D also is a $\mathbb{T}_\infty$-algebra. With these $\mathbb{T}_\infty$-algebra structures, we can state and prove the following strengthening of Proposition I.1.1:

Proposition 4.2. The map $\alpha\colon R[\varepsilon] \to R^D$ is a homomorphism of $\mathbb{T}_\infty$-algebras.

Proof. Let g be any n-ary operation of $\mathbb{T}_\infty$. We should prove commutativity of the diagram (cf. I.(1.4))

$$\begin{array}{ccc} R^n \times R^n \cong (R[\varepsilon])^n & \xrightarrow{\alpha^n} & (R^D)^n \cong (R^n)^D \\ \Big\downarrow{\scriptstyle i(g_{\mathbb{R}[\varepsilon]}) = g_{R[\varepsilon]}} & & \Big\downarrow{\scriptstyle g^D} \\ R[\varepsilon] & \xrightarrow[\alpha]{} & R^D \end{array}$$

Let $(\underline{x},\underline{y}) \in_X R^n \times R^n$. Chasing it the upper way round in this diagram yields the element of R^D

$$g \circ [d \longmapsto (\underline{x}+d\cdot\underline{y})] = [d \longmapsto g(\underline{x}+d\cdot\underline{y})].$$

The g here means the action of the operation g on the algebra $R = i(\mathbb{R})$, so it is really $i(g)$. So let us write that. The right hand side is then, by Taylor's formula (Theorem I.5.2)

$$[d \longmapsto (ig)(\underline{x}) + d\cdot d(ig)_{\underline{x}}(\underline{y})] \quad . \tag{4.4}$$

On the other hand, chasing $(\underline{x},\underline{y})$ the lower way round the diagram gives, because of (4.3)

$$\alpha((ig)(\underline{x}), (i(dg))_{\underline{x}}(\underline{y}))$$

which is

$$[d \longmapsto (ig)(\underline{x}) + d\cdot(i(dg))_{\underline{x}}(\underline{y})] \ . \qquad (4.5)$$

Comparing (4.4) and (4.5), we see that the commutativity follows from $d(ig) = i(dg)$, which was our main "comparison" result of §3 (Theorem 3.3).

Since the α_U of (4.2), for $U = \mathbb{R}^n$, under the identifications $T(R^n) = R^n \times R^n = (R[\varepsilon])^n$ and $(R^n)^D = (R^D)^n$ gets identified with $\alpha^n\colon (R[\varepsilon])^n \longrightarrow (R^D)^n$, we see that the Proposition can be read: the family of maps α_U of (4.2) is natural with respect to smooth maps $\mathbb{R}^n \longrightarrow \mathbb{R}$:, since α_U is also, by construction, natural with respect to the inclusions $U \subseteq \mathbb{R}^n$, we see that the family α_U (for U ranging over open subsets of coordinate vector spaces $\mathbb{R}^n$) is natural with respect to those smooth maps $g\colon U \to \mathbb{R}$ which have smooth extensions to all $\mathbb{R}^n \supseteq U$.

To prove naturality of the α_U (where $U \subseteq \mathbb{R}^n$ is open) with respect to arbitrary smooth mappings $g\colon U \longrightarrow \mathbb{R}$, we use a well known device from smooth analysis: we construct an open covering $\{U_j \subseteq U \mid j \in J\}$ of U such that each restriction $g|U_j$ has a smooth extension to all of $\mathbb{R}^n$. Now the tangent bundle functor $T\colon \mathrm{Mf} \longrightarrow \mathrm{Mf}$ preserves open coverings, and i takes open coverings to coverings, by Axiom C. The naturality assertion we want for α with respect to g is the assertion of equality of two arrows $i(TU) \rightrightarrows R^D$; we have a covering, i.e. a jointly epic family $\{i(TU_j) \to i(TU) \mid j \in J\}$ such that the composites

$$i(TU_j) \longrightarrow i(TU) \rightrightarrows R^D$$

are equal, by the naturality of α with respect to $g|U_j$. This implies the equality of the two maps $i(TU) \to R^D$, and thus the naturality of α with respect to $g\colon U \longrightarrow \mathbb{R}$. It is now immediate to deduce naturality of α with respect to any smooth $U \to \mathbb{R}^m$, and then also with respect to any smooth $U \to V$ with $U \subseteq \mathbb{R}^n$ and $V \subseteq \mathbb{R}^m$ open subsets.

Proposition 5.1. The $\mathbb{T}_\infty$-algebra $C^\infty(\mathbb{R}^n)$ is the free $\mathbb{T}_\infty$-algebra in n generators (namely the $\mathrm{proj}_i\colon \mathbb{R}^n \longrightarrow \mathbb{R}$).

Proof. This is evident from $C^\infty(\mathbb{R}^n) = \mathbb{T}_\infty(n,1)$.

In particular, $C^\infty(\mathbb{R}^n)$ is finitely presented as a $\mathbb{T}_\infty$-algebra.

The main results to be proved in the next three §'s are:

(0) every Weil algebra over $\mathbb{R}$ is canonically a $\mathbb{T}_\infty$-algebra (and is, as such, finitely presented).

(i) the functor $C^\infty(-)\colon Mf \to (\mathbb{T}_\infty\text{-Alg})^{op}$ is full and faithful.

(ii) it preserves transversal pull-backs, and

(iii) it factors through the subcategory $(FP\mathbb{T}_\infty)^{op}$ of finitely presented $\mathbb{T}_\infty$-algebras;

(iv) there is a Grothendieck topology on $(FG\mathbb{T}_\infty)^{op}$ whose restriction to the subcategory $Mf \hookrightarrow (FG\mathbb{T}_\infty)^{op}$ is the standard "open cover topology", and whose "reflection" to a certain subcategory $\mathcal{B}^{op}$ (containing $FP\mathbb{T}_\infty^{op}$) is subcanonical.

The category $\mathcal{B} \subseteq FG\mathbb{T}_\infty$ is the category of "germ-determined $\mathbb{T}_\infty$-algebras", or $\mathbb{T}_\infty$-algebras "presented by an ideal of local character", cf. §6 below; besides the finitely presented $\mathbb{T}_\infty$-algebras, $\mathcal{B}$ also contains for instance the algebra of germs of smooth functions at a point of a manifold.

Having these results, well adapted models can be constructed by algebraic means, like in §1, by taking $\underline{\mathrm{Set}}^{\mathcal{B}}$, or better $\widetilde{\mathcal{B}^{op}}$ with respect to the topology mentioned in (iv); the proof of Axiom $1^W_{\mathbb{R}}$ then goes completely like in §1. But at the same time, we have the axiomatics of §4 available, namely with the composite functor

$$\mathrm{Mf} \xrightarrow{C^{\infty}(-)} \mathcal{B}^{op} \xrightarrow{y} \widetilde{\mathcal{B}^{op}} = \mathcal{E}$$

as our embedding $i: \mathrm{Mf} \to \mathcal{E}$. This essentially follows from the results (i)-(iv) quoted above.

The plan for proving (0)-(iv) is as follows: in the present §5, we describe some general "commutative algebra" of the category $\mathbb{T}_\infty$-Alg and prove (0) and (i). In §6, we introduce the notion 'ideal of local character', or 'germ-determined ideal', and prove that every finitely generated ideal is such. This then leads to a proof of some special cases of (ii). In §7, we introduce a Grothendieck topology on $\mathcal{B}^{op}$. Its good properties lead to a 'patching' procedure through which the general preservation of transversal pullbacks by the functor $C^{\infty}(-)$ is reduced to the special case already studied.

The following facorization property in the theory $\mathbb{T}_\infty$ is essentially obtained by iterating Hadamard's lemma:

Proposition 5.2. Let $\varphi: \mathbb{R}^n \longrightarrow \mathbb{R}$ be smooth (i.e. $\varphi \in \mathbb{T}_\infty(n,1)$), and let $k \geq 0$ be an integer. Then there exists unique $\varphi_\alpha \in \mathbb{T}_\infty(n,1)$ and $\psi_\beta \in \mathbb{T}_\infty(2n,1)$ so that, for all $(\underline{x},\underline{y}) \in \mathbb{R}^n \times \mathbb{R}^n$

$$\varphi(\underline{x}+\underline{y}) = \sum_{|\alpha| \leq k} \varphi_\alpha(\underline{x}) \cdot \underline{y}^\alpha + \sum_{|\beta|=k+1} \psi_\beta(\underline{x},\underline{y}) \cdot \underline{y}^\beta, \tag{5.1}$$

(where α and β denote multi-indices in n letters, and we use the standard conventions for such, quoted in I §5). Note that $\varphi_o = \varphi$.

As a first application, we prove

Theorem 5.3. Let W be a Weil algebra over $\mathbb{R}$, and let $B \in \mathbb{T}_\infty$-Alg. Then there exists a unique $\mathbb{T}_\infty$-algebra structure on $B \otimes_{\mathbb{R}} W$ extending its $\mathbb{R}$-algebra structure, and such that the canonical $j: B \longrightarrow B \otimes_{\mathbb{R}} W$ becomes a $\mathbb{T}_\infty$-homomorphism.

In particular, any Weil algebra over $\mathbb{R}$ carries a canonical $\mathbb{T}_\infty$-algebra structure, and $B \otimes_{\mathbb{R}} W$ is the coproduct of B and W in $\mathbb{T}_\infty$-Alg. With the canonical $\mathbb{T}_\infty$-algebra structure on W
$$\hom_{\mathbb{R}\text{-Alg}}(W,C) = \hom_{\mathbb{T}_\infty\text{-Alg}}(W,C), \quad \text{for any } \mathbb{T}_\infty\text{-algebra } C.$$

Proof. (This is really a straightforward generalization of an argument in §4, cf. (4.3).) W can, as a vector space, be written $W = \mathbb{R} \oplus W'$ with every element in W' nilpotent, in fact, there exists an integer $k \geq 0$ so that the product in W of any $k+1$ elements from W' is 0. Therefore $B \otimes_{\mathbb{R}} W = B \oplus (B \otimes_{\mathbb{R}} W')$, and the product of any $k+1$ elements from $B \otimes_{\mathbb{R}} W'$ is 0. Let $\varphi \in \mathbb{T}_\infty(n,1)$, and let

$$r_i = j(x_i) + y_i \qquad i = 1,\ldots,n$$

be an n-tuple of elements of $B \otimes_{\mathbb{R}} W = B \oplus (B \otimes_{\mathbb{R}} W')$. If $B \otimes_{\mathbb{R}} W$ has a $\mathbb{T}_\infty$-algebra structure, then by the equation (5.1) (which holds for all $\mathbb{T}_\infty$-algebras, since it is an equation in $\mathbb{T}_\infty$) we get

$$\varphi(\underline{r}) = \varphi(j(\underline{x}) + \underline{y}) = \sum_{|\alpha| \leq k} \varphi_\alpha(j(\underline{x})) \cdot \underline{y}^\alpha + \sum_{|\beta| = k+1} \psi_\beta(j(\underline{x}), \underline{y}) \cdot \underline{y}^\beta ,$$

where $\underline{r} = (r_1, \ldots, r_n)$, etc. The β-summation is zero, so that we must have

$$\varphi(\underline{r}) = \sum_{|\alpha| \leq k} \varphi_\alpha(j(\underline{x})) \cdot \underline{y}^\alpha = \sum_{|\alpha| \leq k} j(\varphi_\alpha(\underline{x})) \cdot \underline{y}^\alpha, \qquad (5.2)$$

the last equality sign by the assumption that $j: B \to B \otimes_{\mathbb{R}} W$ is a $\mathbb{T}_\infty$-algebra homomorphism. This proves the uniqueness. The fact that (5.2) actually does define a $\mathbb{T}_\infty$-algebra structure on $B \otimes_{\mathbb{R}} W$ is now straightforward. Taking $B = \mathbb{R}$, we get the canonical $\mathbb{T}_\infty$-algebra structure on $W = \mathbb{R} \otimes_{\mathbb{R}} W$. The proof of the assertion about coproducts as well as the assertion about hom-sets is similar to the proof of the uniqueness of $\mathbb{T}_\infty$-structure just given.

In commutative algebra, we have the three fundamental constructions of 1) forming B/I for I an ideal of B, 2) forming $B[\Sigma^{-1}]$ for Σ a subset of B, and 3) adjoining an indeterminate element to B. Each of these constructions solves a universal problem, namely, respectively of making all elements of I zero, making all elements in Σ invertible, and taking coproduct of B with the free thing in one generator. Furthermore, the two constructions B/I and $B[\Sigma^{-1}]$ 'commute' with each other, in an evident sense. We may ask for the similar constructions in $\mathbb{T}_\infty$-Alg. In so far as the first is concerned, the answer is that the construction is the <u>same</u> as for $\mathbb{R}$-algebras , more precisely:

<u>Proposition 5.4</u>. Let B be a $\mathbb{T}_\infty$-algebra, and let $I \subseteq B$ be an ideal (in the usual ring theoretic sense). Then the $\mathbb{R}$-algebra B/I carries a unique structure of $\mathbb{T}_\infty$-algebra such that the natural $B \longrightarrow B/I$ is a $\mathbb{T}_\infty$-algebra homomorphism.

Equivalently, the relation: $x \sim y$ iff $y-x \in I$, is a congruence relation for the theory $\mathbb{T}_\infty$.

<u>Proof</u>. We prove the last statement. Suppose $y_i - x_i = h_i \in I$, $i = 1,\dots,n$. For any $\varphi \in \mathbb{T}_\infty(n,1)$, we must prove $\varphi(\underline{y}) - \varphi(\underline{x}) \in I$. Now using Proposition 5.2 with $k = 1$, we have

$$\varphi(\underline{y}) = \varphi(\underline{x} + \underline{h}) = \varphi_o(\underline{x}) + \sum_{|\beta|=1} \psi_\beta(\underline{x},\underline{h}) \cdot \underline{h}^\beta \quad .$$

But $\varphi_o = \varphi$, and the sum Σ_β belongs to I since each h_i does.

Because of the Proposition, 'ordinary' ideals play the same role in the theory of $\mathbb{T}_\infty$-algebras as they do in commutative algebra; some titles of famous books reflect this fact, [57], [75], even though these books do not explicitely utilize the $\mathbb{T}_\infty$-algebra viewpoint.

Example 5.5. If $p \in \mathbb{R}^n$, we let $J(p)$ be the ideal of
functions f for which there is an open neighbourhood of p where
f vanishes. The $\mathbb{T}_\infty$-algebra $C^\infty(\mathbb{R}^n)/J(p)$ is the algebra of germs
at p. The ideal $J(p)$ is not finitely generated, so there is no
reason why $C^\infty(\mathbb{R}^n)/J(p)$ should be finitely presented. It is, of
course, finitely generated.

We next consider the construction of inverting the elements
of a subset $\Sigma \subseteq B$, where B is a $\mathbb{T}_\infty$-algebra. Here, the $\mathbb{R}$-alge-
bra $B[\Sigma^{-1}]$ will not in general carry a $\mathbb{T}_\infty$-algebra structure. We
denote by $B \to B\{\Sigma^{-1}\}$ the universal solution in $\mathbb{T}_\infty$-Alg of making
the elements of Σ invertible. Unlike in ordinary commutative alge-
bra, we do not in general have an explicit description of the elements of $B\{\Sigma^{-1}\}$.

Finally, for general reasons, $\mathbb{T}_\infty$-Alg has coproducts, which
we denote $\otimes_\infty$. One cannot describe these as explicitly as in
commutative algebra. But from $C^\infty(\mathbb{R}^n)$ being the free algebra in n
generators follows that

$$C^\infty(\mathbb{R}^n) = C^\infty(\mathbb{R}) \otimes_\infty C^\infty(\mathbb{R}) \otimes_\infty \cdots \otimes_\infty C^\infty(\mathbb{R}) \quad \text{(n times)} \quad (5.3)$$

so that at least this $\otimes_\infty$-formation is easy to describe explicitly.
Adjoining an indeterminate t to the $\mathbb{T}_\infty$-algebra B consists just
in forming $C^\infty(\mathbb{R}) \otimes_\infty B$. - Note that (5.3) may be read as:
$C^\infty(-): Mf \longrightarrow (\mathbb{T}_\infty\text{-Alg})^{op}$ preserves finite products of $\mathbb{R}^m$'s.

For the case where Σ consists of just one element b, the
formation of $B\{\Sigma^{-1}\} = B\{b^{-1}\}$ may be reduced to a combination of
the two other constructions: (i) adjoin an indeterminate t to B,
and (ii) divide out by the ideal generated by $t \cdot b - 1$,

$$B\{b^{-1}\} = (C^\infty(\mathbb{R}) \otimes_\infty B)/(t \cdot b - 1) \quad (5.4)$$

This is immediate from the universal properties involved. In particular, if B is finitely generated (respectively finitely presented) as a $\mathbb{C}_\infty$-algebra, then so is $B\{b^{-1}\}$.

The feature which distinguishes the commutative algebra of $\mathbb{C}_\infty$-Alg from that of $\mathbb{R}$-Alg is the existence of <u>locally finite partitions of unity</u>. Recall (from [72] Coroll. 2 p.6, say) that if $\{U_i \mid i \in I\}$ is an open covering of a manifold M, there exists a <u>locally finite partition of unity subordinate to the covering</u> $\{U_i \mid i \in I\}$, meaning a family $\{\varphi_i \mid i \in I\}$ of smooth functions $\varphi_i : M \to \mathbb{R}$ with

(i) $\mathrm{supp}(\varphi_i) \subseteq U_i$ (recall $\mathrm{supp}(\varphi_i)$ = closure of the set $\{x \in M \mid \varphi_i(x) \neq 0\}$)

(ii) for all $x \in M$, $\exists$ open set $V \ni x$ such that the set $\{i \in I \mid V \cap \mathrm{supp}(\varphi_i) \neq \emptyset\}$ is finite

(iii) $\sum_{i \in I} \varphi_i \equiv 1$

(Note that the sum in (iii) makes sense because of (ii)).

If $\{U_i \mid i \in I\}$ and $\{\varphi_i \mid i \in I\}$ are as above, we let $W_i = \{x \mid \varphi_i(x) \neq 0\}$ (so $\mathrm{supp}\ \varphi_i = \bar{W}_i$, and W_i is open). Let $J \subseteq I$ be the set $\{j \in I \mid W_j \neq \emptyset\}$. Clearly the family $\{W_j \mid j \in J\}$ is a locally finite refinement of $\{U_i \mid i \in I\}$, meaning that it is

(i) an open covering, and that it is

(ii) locally finite (i.e. for all $x \in M$ $\exists$ open set $V \ni x$ such that the set $\{j \in J \mid V \cap \bar{W}_j \neq \emptyset\}$ is finite)

(iii) subordinate to the covering $\{U_i \mid i \in I\}$, i.e., for each $j \in J$, there exists an $i \in I$ with $\bar{W}_j \subseteq U_i$.

As a first application of partition of unity (for an open cover with just two sets), we give two equivalent descriptions of $C_p^\infty(M)$, the $\mathbb{T}_\infty$-algebra of germs at $p \in M$, where M is a manifold. One is an immediate generalization of Example 5.5, namely

$$C_p^\infty(M) := C^\infty(M)/J(p),$$

where $J(p)$ is the ideal of functions vanishing on some open neighbourhood of p (i.e. the ideal of functions whose germ at p is 0). The other is

$$C_p^\infty(M) := C^\infty(M)\{\Sigma_p^{-1}\}$$

where Σ_p is the set of those $f \in C^\infty(M)$ with $f(p) \neq 0$. To prove that these two descriptions are mutually consistent, we should prove

<u>Proposition 5.6</u>. For any $p \in M$, there is a canonical isomorphism

$$C^\infty(M)/J(p) \cong C^\infty(M)\{\Sigma_p^{-1}\}$$

<u>Proof</u>. To produce a map $C^\infty(M)/J(p) \longrightarrow C^\infty(M)\{\Sigma_p^{-1}\}$, it suffices to see that any $f \in J(p)$ becomes zero in $C^\infty(M)\{\Sigma_p^{-1}\}$. Let f vanish on $U \ni p$, and let φ, ψ be a partition of unity subordinate to the covering $U, \neg\{p\}$. Then $\psi(p) = 0$, so $\varphi(p) = 1$, so $\varphi \in \Sigma_p$. But $\varphi \cdot f \equiv 0$ since $\mathrm{supp}(\varphi) \subseteq U$ and f vanishes on U. So f is 0 in $C^\infty(\mathbb{R}^n)\{\Sigma_p^{-1}\}$.

To produce a map the other way, we should prove that any $g \in \Sigma_p$ is invertible mod $J(p)$. If $g(p) \neq 0$, $g \neq 0$ in some neighbourhood $U \ni p$. Let φ, ψ be a partition of unity subordinate to $U, \neg\{p\}$. Then $\varphi \cdot g^{-1}$ is well defined and smooth throughout M (here we use that the support of φ is <u>closed</u> and contained in the open set U); so

$$g \cdot (\varphi \cdot g^{-1}) - 1 = \varphi - 1 = -\psi \ .$$

But ψ (and hence $-\psi$) belongs to $J(p)$; for $\mathrm{supp}(\psi) \subseteq \neg\{p\}$, so $\neg\mathrm{supp}(\psi)$ is an open neighbourhood around p, and ψ vanishes on it.

The element in $C^\infty_p(M)$ determined by $f \in C^\infty(M)$ is called the <u>germ</u> of f <u>at</u> $p \in M$, and denoted f_p. If $I \subseteq C^\infty(M)$ is an ideal, the set $\{f_p \mid f \in I\}$ is an ideal in $C^\infty_p(M)$, denoted I_p.

The following result is classical; it is called 'Milnor's Exercise' in [11], and we refer the reader to there for a proof.

<u>Theorem 5.7</u>. Let M be a manifold. To any $\mathbb{R}$-algebra map $p: C^\infty(M) \longrightarrow \mathbb{R}$, there exists a unique point $P \in M$, such that p is "evaluation at P",

$$p(f) = f(P) \qquad \forall f \in C^\infty(M)$$

Since "evaluation at P" is clearly a Π_∞-algebra homomorphism, the theorem is equally true if we instead of "$\mathbb{R}$-algebra map" write "Π_∞-algebra homomorphism".

<u>Corollary 5.8</u>. The functor $C^\infty(-): \mathrm{Mf} \to (\mathbb{R}\text{-Alg})^{op}$, and hence also the functor $C^\infty(-): \mathrm{Mf} \to (\Pi_\infty\text{-Alg})^{op}$, are full and faithful.

<u>Proof</u>. Let $\varphi: C^\infty(N) \to C^\infty(M)$ be an $\mathbb{R}$-algebra homomorphism. Let $P \in M$, and let $p: C^\infty(M) \longrightarrow \mathbb{R}$ be evaluation at P, as in the theorem. By the theorem, applied for N, the composite $p \circ \varphi$ is "evaluation at $Q \in N$" for some unique $Q \in N$, which we denote $f(P)$. Thus we have defined at least a set-theoretic mapping $f: M \to N$. To prove that f is smooth, it suffices to see that $g \circ f$ is smooth for any smooth $g: N \to \mathbb{R}$. But we claim

$$g \circ f = \varphi(g) \in C^\infty(M), \tag{5.5}$$

from which smoothness follows. To see (5.5), it suffices to see that the two mappings there agree on any $P \in M$. But $\varphi(g)(P) = g(f(P))$, by construction of f. This proves (5.5) and also that $\varphi = C^\infty(f)$. So $C^\infty(-)$, considered as a functor with values in $(\mathbb{R}\text{-Alg})^{op}$, is full and faithful. Since the forgetful functor $(\Pi_\infty\text{-Alg})^{op} \to (\mathbb{R}\text{-Alg})^{op}$ is faithful, it is obvious that $C^\infty(-)$ is also full and faithful when considered to have values in $(\Pi_\infty\text{-Alg})^{op}$.

Because of Theorem 5.7, and the remark following it, it is consistent to formulate

Definition 5.9. Let $B \in \Pi_\infty\text{-Alg}$. A point p of B is a Π_∞-algebra homomorphism $B \to \mathbb{R}$. We say that B is point-determined if for any $f \in B$

$$(p(f) = 0 \text{ for all points } p \text{ of } B) \Rightarrow (f = 0).$$

Corollary 5.10. A Π_∞-algebra of form $C^\infty(M)$ $(M \in Mf)$ is point-determined.

Note that under the bijective correpondence of Theorem 5.7 between points $P \in M$ and points p of $C^\infty(M)$, $f(P) = p(f)$. We shall confuse P and p notationally.

We end this § by some Propositions about Weil algebras (over $\mathbb{R}$) viewed as Π_∞-algebras.

Proposition 5.11. Let $f_i(X_1,\dots,X_n)$ $(i = 1,\dots,k)$ be polynomials with coefficients from $\mathbb{R}$, and let $I \subseteq \mathbb{R}[X_1,\dots,X_n]$ be the ideal they generate. Suppose $\mathbb{R}[X_1,\dots,X_n]/I$ is a Weil algebra W. Let $J \subseteq C^\infty(\mathbb{R}^n)$ be the ideal generated by the functions $f_1,\dots,f_k$. Then the canonical $\mathbb{R}$-algebra map

$$W = \mathbb{R}[X_1,\dots,X_n]/I \longrightarrow C^\infty(\mathbb{R}^n)/J$$

is an isomorphism, in fact an isomorphism of Π_∞-algebras.

In particular, any Weil algebra (over $\mathbb{R}$) is finitely presented as a Π_∞-algebra.

Proof. This is a straightforward application of "Hadamard's lemma" in the form of Proposition 5.2, and for the last assertion, also of Proposition 5.1.

Proposition 5.12. Let W be a Weil algebra over $\mathbb{R}$ and let $\varphi: C^\infty(\mathbb{R}^m) \to W$ be a Π_∞-homomorphism, "centered" at $p \in \mathbb{R}^m$ (meaning: the composite $\pi \circ \varphi: C^\infty(\mathbb{R}^m) \longrightarrow \mathbb{R}$ is "evaluation at $p \in \mathbb{R}^m$"). Then for any linear $\rho: W \to \mathbb{R}$, there exists a polynomial P in m variables such that

$$\forall f \in C^\infty(\mathbb{R}^m): \quad \rho(\varphi(f)) = (P(\frac{\partial}{\partial x_1}, \ldots, \frac{\partial}{\partial x_n})(f))(p).$$

Proof. Any Weil algebra W over $\mathbb{R}$ is a quotient of one of form $C^\infty(\mathbb{R}^n)/I_{k+1}$, where I_{k+1} is the ideal generated by monomials of degree $k+1$; so it suffices to consider such Weil algebras. We consider first a special case, namely where φ is the canonical residue class map $\bar{\varphi}: C^\infty(\mathbb{R}^n) \longrightarrow C^\infty(\mathbb{R}^n)/I_{k+1}$. It is centered at $\underline{0} \in \mathbb{R}^n$. An $\mathbb{R}$-linear basis of $W = C^\infty(\mathbb{R}^n)/I_{k+1}$ consists of the monomials of total degree $\leq k$. For any $f \in C^\infty(\mathbb{R}^n)$, we have, by Hadamard's lemma (Proposition 5.2),

$$f(\underline{x}) = \sum_{|\alpha| \leq k} \frac{1}{\alpha!} \frac{\partial^\alpha f}{\partial x^\alpha}(\underline{0}) \cdot \underline{x}^\alpha + \sum_{|\beta| = k+1} \underline{x}^\beta \cdot g_\beta(\underline{x}),$$

so that the residue class $\bar{\varphi}(f)$ is just the α-part of the sum. If $\{\rho_\alpha\}_{|\alpha| \leq k}$ is the dual basis to the basis $\{\underline{x}^\alpha\}_{|\alpha| \leq k}$ for W, we thus have

$$\rho_\alpha(\bar{\varphi}(f)) = \frac{1}{\alpha!} \frac{\partial^\alpha f}{\partial x^\alpha}(\underline{0}),$$

proving the assertion for the ρ_α's. But these form a basis for the vector space of all linear functionals on W.

To prove the general case, let $\varphi: C^\infty(\mathbb{R}^m) \longrightarrow W$ be arbitrary, centered at $p \in \mathbb{R}^m$. We have now the situation

$$\begin{array}{ccc} C^\infty(\mathbb{R}^m) & & \\ \Big\downarrow{\scriptstyle C^\infty(h)} & \searrow{\scriptstyle \varphi} & \\ C^\infty(\mathbb{R}^n) & \xrightarrow[\bar{\varphi}]{} & W. \end{array} \tag{5.6}$$

Since $C^\infty(\mathbb{R}^m)$ is free and $\bar{\varphi}$ is surjective, φ lifts over $\bar{\varphi}$ (dotted arrow), and since $C^\infty(-): \mathrm{Mf} \longrightarrow (\mathbb{T}^\infty\text{-Alg})^{op}$ is full, the lifted map is of form $C^\infty(h)$ for some smooth $h: \mathbb{R}^n \longrightarrow \mathbb{R}^m$, with $h(\underline{0}) = p$. If now $\rho: W \to \mathbb{R}$ is a linear functional, then, for any $g \in C^\infty(\mathbb{R}^m)$

$$\rho(\varphi(g)) = \rho(\bar{\varphi}(g \circ h)) = P(\frac{\partial}{\partial x_1}, \ldots, \frac{\partial}{\partial x_n})(g \circ h)(\underline{0}),$$

by the special case already proved. But, as is well known,

$$P(\frac{\partial}{\partial x_1}, \ldots, \frac{\partial}{\partial x_n})(g \circ h) = \sum_i (Q_i(\frac{\partial}{\partial y_1}, \ldots, \frac{\partial}{\partial y_m})(g) \circ h) \cdot (R_i(\frac{\partial}{\partial x_1}, \ldots, \frac{\partial}{\partial x_n})(h))$$

for suitable polynomials Q_i and R_i which depend only on P, not on g and h. In particular (since h is fixed, and $h(\underline{0}) = p$),

$$P(\frac{\partial}{\partial x_1}, \ldots, \frac{\partial}{\partial x_n})(g \circ h)(\underline{0}) = \sum_i Q_i(\frac{\partial}{\partial y_1}, \ldots, \frac{\partial}{\partial y_m})(g)(p) \cdot r_i$$

which is of form $Q(\frac{\partial}{\partial y_1}, \ldots, \frac{\partial}{\partial y_m})(g)(p)$ for a polynomial Q. This proves the Proposition.

EXERCISES

5.1. (Dubuc [11]). Let W be a Weil algebra over $\mathbb{R}$, and let B be an arbitrary $\mathbb{T}_\infty$-algebra. Prove

$$\hom_{\mathbb{T}_\infty\text{-Alg}}(B, W) = \hom_{\mathbb{R}\text{-Alg}}(B, W).$$

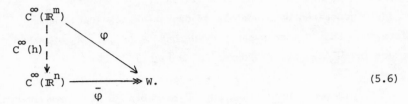

5.2. Show that the ring of formal power series $\mathbb{R}[[X_1,\ldots,X_n]]$ carries a canonical structure of $\mathbb{T}_\infty$-algebra (Hint: it is an inverse limit of Weil algebras $\mathbb{R}[X_1,\ldots,X_n]/I_{k+1}$, where I_{k+1} is the ideal generated by monomials of degree k+1.) (Dubuc has informed me that Calderon has proved that $\mathbb{R}[[X_1,\ldots,X_n]]$ is the coproduct in $\mathbb{T}_\infty$-Alg of n copies of $\mathbb{R}[[X]]$.)

5.3 (Reyes). Prove that any $\mathbb{T}_\infty$-algebra B is formally-real in the sense that, for any n = 1,2,..., and any $b_1,\ldots,b_n \in B$, $1+\Sigma(b_i^2)$ is invertible in B. (Hint: we have a smooth function $\mathbb{R}^n \longrightarrow \mathbb{R}$ given by $(x_1,\ldots,x_n) \longmapsto (1+\Sigma(x_i^2))^{-1}$; interpret it as an n-ary operation in $\mathbb{T}_\infty$.)

5.4. Let M be a manifold, and $F \subseteq M$ a subset. Let $I_F \subseteq C^\infty(M)$ be the ideal of functions that vanish on F. For I an ideal of $C^\infty(M)$, let $Z(I) \subseteq M$ be the set

$$Z(I) = \{x \in M \mid f(x) = 0 \ \ \forall f \in I\} \tag{5.7}$$

Prove that $F = Z(I_F)$ if and only if F is closed. (Hint: Use the existence of smooth characteristic functions for open sets of a manifold, cf. e.g. proof of Theorem 3.4).

5.6. With notation as in Exercise 5.5, prove that for an ideal $I \subseteq C^\infty(M)$, we have $I = I_{Z(I)}$ if and only if $C^\infty(M)/I$ is point-determined.

5.7 (Dubuc and Schanuel). Let $I \subseteq C^\infty(\mathbb{R}^m)$ be an ideal. Let I^* denote the ideal generated by the image of I under the inclusion $C^\infty(\mathbb{R}^m) \longrightarrow C^\infty(\mathbb{R}) \otimes_\infty C^\infty(\mathbb{R}^m) = C^\infty(\mathbb{R}^{m+1})$. Writing elements of this algebra in form $f(t,\underline{x})$, $(t \in \mathbb{R},\ \underline{x} \in \mathbb{R}^m)$, prove that

$$t\cdot f(t,\underline{x}) \in I^* \Rightarrow f(t,\underline{x}) \in I^* \quad .$$

III.6: GERM-DETERMINED Π_∞-ALGEBRAS

It is now consistent to formulate

Definition 6.1. Let $B \in \Pi_\infty$-Alg, and let p be a point of B, $p: B \to \mathbb{R}$. The germ-algebra at p, denoted B_p, is the Π_∞-algebra $B\{\Sigma_p^{-1}\}$ where $\Sigma_p = \{f \in B \mid p(f) \neq 0\}$.

If $f \in B$, the element it represents in B_p is denoted f_p, and called the germ of f at the point p. Similarly, if I is an ideal in B, the smallest ideal in B_p containing $\{f_p \mid f \in I\}$ is denoted I_p.

If $M \in \mathrm{Mf}$, $B = C^\infty(M)$, and $P \in M$, and if $p: C^\infty(M) \longrightarrow \mathbb{R}$ is the point (in the sense of Definition 5.9) defined as "evaluation at P", then $p(f) = f(P)$, and f_p as defined just now agrees with the germ (in the classical sense) of f at P, by Proposition 5.6. Confusing P and p notationally, we may identify $C_p^\infty(M)$ (the classical ring of germs at p) with $(C^\infty(M))_p$, as defined now.

Definition 6.2. Let $B \in \Pi_\infty$-Alg, and let $I \subseteq B$ be an ideal. The germ-radical of I denoted $\hat{I}$, is the ideal in B given by

$$\hat{I} = \{f \in B \mid f_p \in I_p \text{ for all points } p \text{ of } B\}.$$

Clearly $I \subseteq \hat{I}$. We say that I is of local character, or germ-determined if $I = \hat{I}$. We say that B is germ-determined if the zero ideal in B is germ-determined.

Thus, $I \subseteq B$ is of local character if a sufficient condition for f to belong to I is that $f_p \in I_p$ for all points p of B.

Any Π_∞-algebra B evidently maps to a germ-determined one in a universal way, namely by $B \to B/\{0\}^\wedge$). We may denote $B/(\{0\}^\wedge)$ simply by $\hat{B}$. Note that B and $\hat{B}$ have the same points.

A Π_∞-algebra need not have any points: let $I \subseteq C^\infty(\mathbb{R})$ be the ideal of functions f with compact support. Then for each $p \in \mathbb{R}$, $I_p = C^\infty_p(\mathbb{R})$, so $\hat{I} = C^\infty(\mathbb{R})$; thus, for $B = C^\infty(\mathbb{R})/I$, $\hat{B} = \{0\}$, which has no points, so B has no points. Also the ideal I is not of local character, and B is not germ-determined.

The contention is that the only geometrically interesting Π_∞-algebras are the germ-determined ones. Clearly any point-determined Π_∞-algebra (Definition 5.9), and in particular any $C^\infty(M)$ for $M \in \mathrm{Mf}$, is germ determined. There are many interesting Π_∞-algebras which are germ-determined, but not point-determined, like $\mathbb{R}[\varepsilon]$ (or any other Weil-algebra), or $C^\infty_p(M)$.

We let $\mathcal{B}$ denote the full subcategory of $FG\Pi_\infty$ consisting of germ determined Π_∞-algebras. If $C^\infty(\mathbb{R}^n)/I$ is a presentation of a germ-determined Π_∞-algebra,

$$C^\infty(\mathbb{R}^n)/I = (C^\infty(\mathbb{R}^n)/I)^\wedge = C^\infty(\mathbb{R}^n)/\hat{I} \quad ,$$

so that $I = \hat{I}$; so I is germ-determined. Conversely, if $I \subseteq C^\infty(\mathbb{R}^n)$ is germ-determined, $C^\infty(\mathbb{R}^n)/I$ is germ determined, and belongs to $\mathcal{B}$.

The process $B \longmapsto \hat{B}$ defines a reflection functor $FG\Pi_\infty \longrightarrow \mathcal{B}$, left adjoint to the inclusion $\mathcal{B} \hookrightarrow FG\Pi_\infty$. The existence of that implies that $\mathcal{B}$ has finite colimits. They are not in general preserved by the inclusion functor $\mathcal{B} \hookrightarrow FG\Pi_\infty$.

Let $M \in \mathrm{Mf}$. From Proposition 5.6 follows that if $f \in C^\infty(M)$ and $I \subseteq C^\infty(M)$ is an ideal, then $f \in \hat{I}$ implies that for each $p \in M$, there exists an open neighbourhood $V^{(p)}$ around p, and functions $g^{(p)}$ and $\gamma^{(p)} \in C^\infty(M)$ such that

$$f = g^{(p)} + \gamma^{(p)} \tag{6.1}$$

with $g^{(p)} \in I$ and $\gamma^{(p)}$ vanishing on $V^{(p)}$. (Conversely, an f of this form for all $p \in M$ belongs to $\hat{I}$.) This description is used in the proof of the following essential

<u>Theorem 6.3</u>. Let M be a manifold, $J \subseteq C^{\infty}(M)$ an ideal, and $h \in C^{\infty}(M)$ an arbitrary element. If J is germ-determined, then so is the ideal $I = (J,h)$ generated by J and h.

In particular, any finitely generated ideal in $C^{\infty}(M)$ is germ-determined; and any finitely presented $\mathbb{T}_{\infty}$-algebra is germ-determined.

<u>Proof</u>. Assume $f_p \in I_p$ $\forall p \in M$. So for every $p \in M$, we may find $g^{(p)}$, $\gamma^{(p)}$, and $V^{(p)}$, as in (6.1). Let $\{U_{\alpha} | \alpha \in A\}$ be a locally finite refinement of the covering $\{V^{(p)} \mid p \in M\}$ of M, and let $\{\varphi_{\alpha} \mid \alpha \in A\}$ be a (locally finite) partition of unity subordinate to $\{U_{\alpha} | \alpha \in A\}$. For each α, we have $U_{\alpha} \subseteq V^{(p)}$ for some p. For such p

$$\varphi_{\alpha} \cdot f = \varphi_{\alpha} \cdot g^{(p)} + \varphi_{\alpha} \cdot \gamma^{(p)} = \varphi_{\alpha} \cdot g^{(p)},$$

the last equality sign because $\mathrm{supp}(\varphi_{\alpha}) \subseteq U_{\alpha} \subseteq V^{(p)}$, and $\gamma^{(p)}$ vanishes here. Since $g^{(p)} \in I$, $\varphi_{\alpha} \cdot f \in I$. This holds for each $\alpha \in A$. So we may write

$$\varphi_{\alpha} \cdot f = j_{\alpha} + k_{\alpha} \cdot h \tag{6.2}$$

for suitable $j_{\alpha} \in J$ and $k_{\alpha} \in C^{\infty}(M)$. Now since $\mathrm{supp}(\varphi_{\alpha}) \subseteq U_{\alpha}$ is closed, we have a partition of unity ρ_{α}, $1-\rho_{\alpha}$ subordinate to the covering U_{α}, $\neg\mathrm{supp}(\varphi_{\alpha})$. In particular, $\rho_{\alpha} \equiv 1$ on $\mathrm{supp}(\varphi_{\alpha})$. So $\rho_{\alpha} \cdot \varphi_{\alpha} = \varphi_{\alpha}$. So we have

$$\varphi_{\alpha} \cdot f = \rho_{\alpha} \cdot \varphi_{\alpha} \cdot f = \rho_{\alpha} \cdot j_{\alpha} + (\rho_{\alpha} \cdot k_{\alpha}) \cdot h .$$

We change notation and write j_α for $\rho_\alpha \cdot j_\alpha$ and k_α for $\rho_\alpha \cdot k_\alpha$; so (6.2) holds again, but now with $\text{supp}(j_\alpha) \subseteq U_\alpha$, $\text{supp}(k_\alpha) \subseteq U_\alpha$. So we have

$$f = (\Sigma\varphi_\alpha)\cdot f = \Sigma(\varphi_\alpha \cdot f) = \Sigma(j_\alpha + k_\alpha \cdot h)$$

$$= \Sigma j_\alpha + \Sigma(k_\alpha \cdot h);$$

this last interchange of the order of summation is permissible, since the families $\{j_\alpha\}$ and $\{k_\alpha\}$ are locally finite because the covering $\{U_\alpha\}$ is locally finite. For the same reason, we may re-write $\Sigma(k_\alpha \cdot h)$ into $(\Sigma k_\alpha)\cdot h$. So altogether we get

$$f = \Sigma j_\alpha + (\Sigma k_\alpha)\cdot h . \tag{6.3}$$

Now again by the local finiteness of $\{U_\alpha\}$, the germ $(\Sigma j_\alpha)_p$ is a finite sum of germs $(j_\alpha)_p$, so $(\Sigma j_\alpha)_p \in J_p$, for each $p \in M$. Since J was assumed to be germ-determined, we conclude $\Sigma j_\alpha \in J$, and clearly $(\Sigma k_\alpha)\cdot h \in (h)$. Thus (6.3) proves $f \in (J,h) = I$, as desired. This proves the general part of the theorem. The particular case follows by induction, starting from the observation that $C^\infty(M)$ is germ-determined, thus the zero ideal in it is germ determined. Taking $M = \mathbb{R}^m$ yields the assertion for $\amalg_\infty$-algebras of form $C^\infty(\mathbb{R}^m)/(f_1,\ldots,f_n)$, i.e. for the finitely presented ones.

The following Proposition is an application of the local character (= germ determinedness) of finitely generated ideals in $C^\infty(M)$. It provides a first partial answer to the question: to what extent does $C^\infty(-)\colon Mf \to (\amalg_\infty\text{-Alg})^{op}$ preserve transversal pull-backs ? M denotes a manifold.

Proposition 6.4. Let $h\colon M \to \mathbb{R}^s$ have $\underline{x}$ as regular value. Then the (transversal) pull-back

$$\begin{array}{ccc} H & \hookrightarrow & M \\ \downarrow & & \downarrow h \\ 1 & \xrightarrow[\ulcorner \underline{x} \urcorner]{} & \mathbb{R}^s \end{array}$$

is preserved by the functor $C^\infty(-)\colon Mf \to (\mathbb{T}_\infty\text{-Alg})^{op}$. Equivalently, the equalizer in Mf

$$H \hookrightarrow M \underset{\underline{x}}{\overset{h}{\rightrightarrows}} \mathbb{R}^s$$

is preserved by $C^\infty(-)$ (where $\underline{x}$ denotes the constant map with value $\underline{x}$). In particular, if $M = \mathbb{R}^n$, $C^\infty(H)$ is finitely presented as a $\mathbb{T}_\infty$-algebra.

Proof. The equivalence of the two formulations is a trivial diagrammatic fact, knowing that $C^\infty(1) = \mathbb{R}$ is the terminal object in $(\mathbb{T}_\infty\text{-Alg})^{op}$. So we prove the second formulation.

Without loss of generality, we may assume $\underline{x} = \underline{0} \in \mathbb{R}^s$. If $h = (h_1,\ldots,h_s)$, then H is the submanifold cut out by the s functions $h_1,\ldots,h_s$ (i.e. the meet of their zero sets). We must prove that

$$C^\infty(H) \xleftarrow{\ r\ } C^\infty(M) \underset{0}{\overset{C^\infty(h)}{\leftleftarrows}} C^\infty(\mathbb{R}^s)$$

is a coequalizer in $\mathbb{T}_\infty$-Alg, where r is the restriction map. Since $C^\infty(\mathbb{R}^s)$ is the free $\mathbb{T}_\infty$-algebra in s generators, this is equivalent to proving that

$$C^\infty(M)/(h_1,\ldots,h_s) \cong C^\infty(H), \qquad (6.4)$$

via the restriction map r. Let $f \in C^\infty(M)$ have $r(f) = 0$, i.e. $f \equiv 0$ on H. We must prove that $f \in (h_1,\ldots,h_s)$. Since this ideal is germ-determined (Theorem 6.3) it suffices to prove, for each $p \in M$,

$$f_p \in (h_1, \ldots, h_s)_p \, . \tag{6.5}$$

For $p \notin H$, $(h_1, \ldots, h_s)_p = C^\infty(M)_p$, so (6.5) is automatic. So consider a $p \in H$. By the implicit function theorem and the fact that the h_i's have linearly independent differentials at p (because $h(p) = \underline{0}$ and $\underline{0}$ was assumed a regular value for h), we may choose a smooth coordinate frame around p and $f(p)$ so that, locally around $p = \underline{0}$, $h: \mathbb{R}^n \longrightarrow \mathbb{R}^s$ is projection onto the s first coordinates (it suffices to consider things locally, since (6.5) is a statement about germs). Writing $\mathbb{R}^n = \mathbb{R}^s \oplus \mathbb{R}^{n-s}$, we thus need:

<u>Lemma</u>. Let $f: U \to \mathbb{R}$ vanish on $U \cap \mathbb{R}^{n-s}$ (U an open neighbourhood of $\underline{0} \in \mathbb{R}^s \oplus \mathbb{R}^{n-s}$. Then

$$f = \sum_{i=1}^{s} g_i \cdot x_i$$

for suitable smooth $g_i: U \to \mathbb{R}$ (here, x_i denotes, as usual, projection $\mathbb{R}^n \longrightarrow \mathbb{R}$ to the i'th factor).

<u>Proof</u>. This is another special case of Hadamard's lemma (cf. Proposition 5.2).

We have now identified the kernel of the restriction map $r: C^\infty(M) \longrightarrow C^\infty(H)$ as being exactly the ideal $(h_1, \ldots, h_s)$. To prove that r is surjective means proving that any smooth $f: H \to \mathbb{R}$ may be extended over $H \hookrightarrow M$. Locally, this can be done, because locally H looks like $\mathbb{R}^{n-s} \subseteq \mathbb{R}^n$. To get a global extension, we may apply a partition of unity. This proves (6.4) and thus the Proposition.

Let $U \subseteq M$ be an open subset of a manifold. Recall (§3) that we may find a smooth characteristic function $g: M \to \mathbb{R}$ for it (i.e. $U = \{x \in M \mid g(x) \neq 0\}$). Then $1 \in \mathbb{R}$ is a regular value for the map $h: \mathbb{R} \times M \to \mathbb{R}$ given by $(t,x) \longmapsto t \cdot g(x)$. From Proposition 6.4 follows that the equalizer in Mf

$$U \xrightarrow{\ e\ } \mathbb{R} \times M \underset{1}{\overset{h}{\rightrightarrows}} \mathbb{R}$$

where $e(x) = (g(x)^{-1}, x)$, is preserved by $C^{\infty}(-)$, or equivalently that

$$C^{\infty}(U) \cong C^{\infty}(\mathbb{R} \times M)/(h-1) \tag{6.6}$$

where $h(t,x) = t \cdot g(x)$. Equivalently, by (5.4),

$$C^{\infty}(U) = C^{\infty}(M)\{g^{-1}\}. \tag{6.7}$$

In particular, taking $M = \mathbb{R}^n$, we have $C^{\infty}(\mathbb{R} \times M) = C^{\infty}(\mathbb{R}^{n+1})$, the free $\mathbb{T}_{\infty}$-algebra in $n+1$ generators, so that (6.6) provides a finite presentation of $C^{\infty}(U)$; so we have

Lemma 6.5. If $U \subseteq \mathbb{R}^n$ is open, $C^{\infty}(U)$ is finitely presented as a $\mathbb{T}_{\infty}$-algebra.

We can now prove

Theorem 6.6. For any manifold M, $C^{\infty}(M)$ is finitely presented as a $\mathbb{T}_{\infty}$-algebra.

Proof. By the version of Whitney's embedding theorem also utilized in §4, M is a smooth retract of some open $U \subseteq \mathbb{R}^n$. It follows that $C^{\infty}(M)$ is a retract in $\mathbb{T}_{\infty}$-Alg of $C^{\infty}(U)$, which is finitely presented, by Lemma 6.5. It is standard universal algebra that a retract of a finitely presented thing is finitely presented.

Using (6.7), it is easy to prove:

<u>Proposition 6.7</u>. Let $U \hookrightarrow M$ be an open subset of a manifold, and let $f\colon N \to M$ be a smooth map between manifolds. Then the (transversal) pull-back defining $f^{-1}(U) \subseteq N$ is preserved by the functor $C^\infty(-)\colon Mf \longrightarrow \mathbb{T}_\infty\text{-Alg}$.

<u>Proof</u>. Let g be a smooth characteristic function for U. Then $g \circ f$ is a smooth characteristic function for $f^{-1}(U)$. From the universal properties involved follows that $C^\infty(N)\{(g \circ f)^{-1}\}$ is a pushout of $C^\infty(M)\{g^{-1}\}$ along $C^\infty(f)$. The result follows from (6.7).

<u>EXERCISES</u>

6.1 (Dubuc). Consider the two germ-determined $\mathbb{T}_\infty$-algebras $C^\infty_0(\mathbb{R})$ and $C^\infty(\mathbb{R})$ (the first one is the algebra of germs at $0 \in \mathbb{R}$). Prove that their coproduct in $\mathbb{T}_\infty$-Alg

$$C^\infty_0(\mathbb{R}) \otimes_\infty C^\infty(\mathbb{R}) \qquad (6.8)$$

is isomorphic to $C^\infty(\mathbb{R}^2)/J$, where J is the ideal of functions vanishing in some subset of $\mathbb{R}^2$ of form $]-\varepsilon,\varepsilon[\times \mathbb{R}$ (a "strip around the y-axis"). Prove that J_p is the unit ideal for any p that does not belong to the y-axis. Thus, if $f\colon \mathbb{R}^2 \to \mathbb{R}$ has $f_p = 0$ for all $p \in$ y-axis, $f \in \hat{J}$. In particular, we conclude that the ideal J is not germ-determined, and that (6.8) is not in $\mathcal{B}$.

6.2. Let W be a Weil algebra over $\mathbb{R}$, and $p\colon W \to \mathbb{R}$ its unique point. Prove that the germ algebra W_p equals W, and conclude in particular that W is germ determined.

III.7: THE OPEN COVER TOPOLOGY

By Corollary 5.8, Theorem 6.6, and Theorem 6.3, we have full inclusions

$$\mathrm{Mf} \hookrightarrow (\mathrm{FP}\Pi_{\infty})^{\mathrm{op}} \hookrightarrow \mathcal{B}^{\mathrm{op}} ,$$

(where $\mathcal{B}$ = category of finitely generated germ-determined Π_{∞}-algebras). We want to extend the Grothendieck topology on Mf, given by the open coverings, to a Grothendieck topology on $\mathcal{B}^{\mathrm{op}}$. The definition is almost immediate, when we take the notion of 'point' serious, and recall, (6.7), that for $U \subseteq M$ open, $C^{\infty}(U) \cong C^{\infty}(M)\{g^{-1}\}$ for some $g \in C^{\infty}(M)$, and conversely, for any $g \in C^{\infty}(M)$, $C^{\infty}(M)\{g^{-1}\} = C^{\infty}(U)$ where $U = \{x \in M | g(x) \neq 0\}$. This leads us to utilize the following terminology, which properly speaking refers to $\mathcal{B}^{\mathrm{op}}$ rather than to $\mathcal{B}$:

<u>Definition 7.1</u>. A map $B \to C$ in $\mathcal{B}$ is an <u>open inclusion</u> if it is of form $B \longrightarrow B\{b^{-1}\} \longrightarrow B\{b^{-1}\}^{\wedge}$ for some $b \in B$, and where the two displayed maps are the canonical ones.

It seems that we cannot from "B is germ-determined" conclude "$B\{b^{-1}\}$ is germ-determined", in general, whence we have to make it so by applying the reflection functor $^{\wedge}: \mathrm{FG}\Pi_{\infty} \longrightarrow \mathcal{B}$. However, if B is finitely presented, then so is $B\{b^{-1}\}$, by (5.4), so is already germ-determined, by Theorem 6.3. This in particular applies to $B = C^{\infty}(M)$ by Theorem 6.6, and by the above remarks, it follows that the inclusion $\mathrm{Mf} \hookrightarrow \mathcal{B}^{\mathrm{op}}$ preserves and reflects the notion of open inclusion, in an evident sense: $C^{\infty}(M) \longrightarrow C$ is an open inclusion iff it is of form $C^{\infty}(M) \longrightarrow C^{\infty}(U)$ for some open subset $U \subseteq M$.

We now describe a Grothendieck topology on $\mathcal{B}^{op}$, by describing a pretopology. We work with the dual category $\mathcal{B}$, so that we must describe co-coverings instead of coverings.

Definition 7.2. An open co-cover of a $B \in \mathcal{B}$ is a family $\{\xi_\alpha : B \to B_\alpha \mid \alpha \in A\}$ of open inclusions in $\mathcal{B}$ (Definition 7.1) such that any point $B \longrightarrow \mathbb{R}$ factors through some ξ_α ("Every point of B is a point of some B_α", or "the family is jointly surjective in so far as points are concerned".)

To check that this is a pre-(co-)-topology: the point-surjectivity considition is evidently preserved under composition of covers, and using only the universal property of pushouts in $\mathcal{B}$, it is likewise easily seen to be preserved under pushout. So we need only see

Proposition 7.3. 1) The pushout in $\mathcal{B}$ of an open inclusion is of an open inclusion 2) The composite of two open inclusions is an open inclusion.

Proof. It is clear that $B\{b^{-1}\}^{\wedge}$ solves in $\mathcal{B}$ the universal problem of making $b \in B$ invertible. From this, and the universal property of pushouts, 1) follows. To prove 2), let $c \in B\{b^{-1}\}^{\wedge}$. We should prove that

$$B \longrightarrow (B\{b^{-1}\}^{\wedge})\{c^{-1}\}^{\wedge}$$

is an open inclusion. Consider a presentation of B, $\pi\colon C^\infty(\mathbb{R}^n) \longrightarrow\!\!\!\!\rightarrow B$ with kernel I. We find a $g \in C^\infty(\mathbb{R}^n)$ mapping to b by π, and by the fact that the processes of dividing out by ideals and forming fraction-Π_∞-algebras commute, we have a commutative square (ignore the α's which are for later reference)

$$\begin{array}{ccc} C^{\infty}(\mathbb{R}^n) & \longrightarrow & C^{\infty}(\mathbb{R}^n)\{g_{\alpha}^{-1}\} = C(U_{\alpha}) \\ {\scriptstyle \pi}\big\downarrow & & \big\downarrow{\scriptstyle q_{\alpha}} \\ B & \longrightarrow & B\{b_{\alpha}^{-1}\} \end{array} \tag{7.1}$$

(with $U = \{x \mid g(x) \neq 0\}$ an open subset of $\mathbb{R}^n$). By the universal properties of the constructions involved, this is a pushout in $FG\Pi_{\infty}$. If we apply the reflection functor $^{\wedge}: FG\Pi_{\infty} \longrightarrow \mathcal{B}$ to it, we get a pushout in $\mathcal{B}$, and it looks like (7.1) except that $B\{b^{-1}\}$ is replaced by $B\{b^{-1}\}^{\wedge}$ (the other three objects being already in $\mathcal{B}$). Obviously, $\hat{q}$ is surjective, so we find some $h \in C^{\infty}(U)$ going to c by $\hat{q}$. Let $V = \{x \in U \mid h(x) \neq 0\}$. It is clearly open in $U \subseteq \mathbb{R}^n$, so it is an open subset of $\mathbb{R}^n$. Let $k: \mathbb{R}^n \longrightarrow \mathbb{R}$ be a smooth characteristic function for it. In $\mathcal{B}$, we have the two adjacent pushout diagrams

$$\begin{array}{ccccc} C^{\infty}(\mathbb{R}^n) & \longrightarrow & C^{\infty}(U) & \longrightarrow & C^{\infty}(U)\{h^{-1}\} = C^{\infty}(V) \\ {\scriptstyle \pi}\big\downarrow & & \big\downarrow & & \big\downarrow \\ B & \longrightarrow & B\{b^{-1}\}^{\wedge} & \longrightarrow & (B\{b^{-1}\}^{\wedge})\{c^{-1}\}^{\wedge}. \end{array} \tag{7.2}$$

But we also have a pushout diagram in $\mathcal{B}$

$$\begin{array}{ccc} C^{\infty}(\mathbb{R}^n) & \longrightarrow & C^{\infty}(\mathbb{R}^n)\{k^{-1}\} = C^{\infty}(V) \\ {\scriptstyle \pi}\big\downarrow & & \big\downarrow \\ B & \longrightarrow & B\{\pi(k)^{-1}\}^{\wedge}, \end{array}$$

and comparing it with the total diagram in (7.2), which must be isomorphic to it, we conclude that the composite map in the bottom line of (7.2) is an open inclusion. This proves the Proposition.

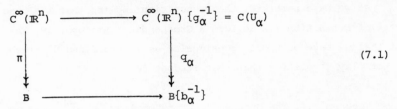

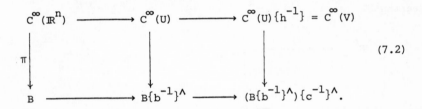

As we have remarked, the Proposition implies that the open covers of Definition 7.2 do form a Grothendieck pretopology. The fact that we have been only considering germ-determined $\mathbb{T}_\infty$-algebras did not play a role for that, but it does for

Theorem 7.4. The Grothendieck pretopology on $\mathcal{B}^{op}$ (as described in Definition 7.2) is subcanonical, i.e. for every $F \in \mathcal{B}$, the functor

$$\hom_{\mathcal{B}}(F,-): \quad (\mathcal{B}^{op})^{op} \longrightarrow \underline{\text{Sets}}$$

is a sheaf.

Proof. Let $B \in \mathcal{B}$, and consider an arbitrary open covering $\{\xi_\alpha: B \to C_\alpha \mid \alpha \in A\}$ of it. As in the proof of Proposition 7.3, it follows that we may choose a presentation $\pi: C^\infty(\mathbb{R}^n) \twoheadrightarrow B$ of B, with kernel I, say, and open subsets U_α of $\mathbb{R}^n$ such that the diagrams

$$\begin{array}{ccc} C^\infty(\mathbb{R}^n) & \longrightarrow & C^\infty(U_\alpha) \\ \pi \big\downarrow & & \big\downarrow q_\alpha \\ B & \longrightarrow & C_\alpha \end{array}$$

are pushouts in $\mathcal{B}$. From the point-surjectivity criterion follows that the union of the U_α's, which we denote W, contains $Z(I) = \{p \in \mathbb{R}^n \mid f(p) = 0 \quad \forall f \in I\}$.

Lemma. The map $\pi: C^\infty(\mathbb{R}^n) \to B$ factors across the restriction map $C^\infty(\mathbb{R}^n) \longrightarrow C^\infty(W)$.

Proof. Let $g: \mathbb{R}^n \longrightarrow \mathbb{R}$ be a smooth characteristic function for W, so $C^\infty(W) = C^\infty(\mathbb{R}^n)\{g^{-1}\}$. We have to see that $\pi(g)$ is invertible in $B = C^\infty(\mathbb{R}^n)/I$. Since B is assumed germ-determined, the ideal $(\pi(g)) \subseteq B$ is germ-determined, by Theorem 6.3. To prove

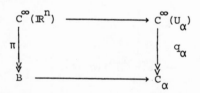

$(\pi(g)) = B$, it suffices to see $(\pi(g))_p = B_p$ for all points p of B, i.e. for all points $p \in Z(I)$. But g is nowhere zero on $Z(I)$, so $p(\pi(g)) \neq 0$ in $\mathbb{R}$, whence $\pi(g)_p$ is invertible in B_p, for $p \in Z(I)$.

By the Lemma, we get a map $\bar{\pi}: C^\infty(W) \longrightarrow B$. We now prove the sheaf condition simultaneously for all $\hom_{\mathcal{B}}(F,-)$, $F \in \mathcal{B}$: let $C_{\alpha\beta}$ denote the pushout in $\mathcal{B}$ of $B \to C_\alpha$ and $B \to C_\beta$. It suffices to prove that

$$B \longrightarrow \Pi C_\alpha \rightrightarrows \Pi C_{\alpha\beta}$$

is an equalizer in $\mathbb{T}_\infty$-Alg. Let $\{\bar{h}_\alpha \in C_\alpha \mid \alpha \in A\}$ be an element in the middle, equalized by the two parallel maps, i.e. $\{\bar{h}_\alpha\}$ is a compatible family. This implies that for all points p of $C_{\alpha\beta}$, $(\bar{h}_\alpha)_p = (\bar{h}_\beta)_p$. Thus, for every point p of B, there is a well-defined element $\bar{h}^{(p)} \in B_p$ with

$$\bar{h}^{(p)} = (\bar{h}_\alpha)_p \tag{7.3}$$

for any α such that p is a point of C_α. We shall construct an element $\bar{h} \in B$ such that

$$\bar{h}_p = \bar{h}^{(p)} \tag{7.4}$$

for all points p of B.

Let $h_\alpha \in C^\infty(U_\alpha)$ have $q_\alpha(h_\alpha) = \bar{h}_\alpha$. Let $\{\varphi_\alpha \mid \alpha \in A\}$ be a locally finite partition of unity of W, subordinate to the covering $\{U_\alpha \mid \alpha \in A\}$ of W. Since $\mathrm{supp}(\varphi_\alpha) \subseteq U$ is closed, the function $\varphi_\alpha \cdot h_\alpha$ is defined and smooth on the whole of W, and, by local finiteness, we may form

$$h := \sum_\alpha \varphi_\alpha \cdot h_\alpha \in C^\infty(W) .$$

We let $\bar{h} := \bar{\pi}(h) \in B$. Generally, in the following, an overbar indicates that we have applied $\bar{\pi}$ or q_α. Now let p be any point of B, (so also $p \in Z(I) \subseteq W$). We choose, by local finiteness of the partition $\{\varphi_\alpha\}$, an open neighbourhood V around p meeting only finitely many $\text{supp}(\varphi_\alpha)$, say those with index in $A_o \subseteq A$. Let $A_1 = \{\alpha \in A_o \mid p \in U_\alpha\}$. We then have

$$\begin{aligned}
(\bar{h})_p &= \sum_{\alpha \in A_1} (\overline{\varphi_\alpha \cdot h_\alpha})_p \\
&= \sum_{A_1} (\bar{\varphi}_\alpha)_p \, (\bar{h}_\alpha)_p \\
&= (\sum_{A_1} (\bar{\varphi}_\alpha)_p) \cdot \bar{h}^{(p)} \qquad \text{(by (7.3))} \\
&= (\overline{\sum_{A_1} \varphi_\alpha})_p \cdot \bar{h}^{(p)} \\
&= \bar{h}^{(p)} ,
\end{aligned}$$

the last equality sign by observing that in a small enough neighbourhood around p, $\sum_{A_1} \varphi_\alpha = \sum_A \varphi_\alpha (\equiv 1)$. This proves (7.4). But then i C_α,

$$\xi_\alpha(\bar{h}) = \bar{h}_\alpha \ ;$$

for, since C_α is germ determined, this follows by proving

$$(\xi_\alpha(\bar{h}))_p = (\bar{h}_\alpha)_p \quad \text{for all points } p \text{ of } C_\alpha;$$

and by (7.3) and (7.4)

$$(\xi_\alpha(\bar{h}))_p = \bar{h}_p = \bar{h}^{(p)} = (\bar{h}_\alpha)_p$$

This proves the Theorem.

EXERCISES

7.1. Prove that in $\mathcal{B}$, the only object covered by the empty family is the $\mathbb{T}_\infty$-algebra $\{0\}$ (the one-element $\mathbb{T}_\infty$-algebra).

7.2. Describe a Grothendieck pre-co-topology on $FG\mathbb{T}_\infty$ by letting an open co-cover of $B \in FG\mathbb{T}_\infty$ be a family $\{\xi_\alpha: B \to B_\alpha | \alpha \in A\}$ which is surjective in so far as points are concerned, and where each ξ_α is an "open pre-inclusion", where this latter just means a map in $FG\mathbb{T}_\infty$ of form $B \to B\{b^{-1}\}$. (Hint: to prove the axioms for pretopologies, modify the proof of Proposition 7.3.)

7.3 (Dubuc). Let $B = C^\infty(\mathbb{R})/I$ where I is the ideal of functions of compact support. In the Grothendieck co-topology of Exercise 7.2, prove that B is covered by the empty family; and conclude that this co-topology on $FG\mathbb{T}_\infty$ is not subcanonical.

III.8: CONSTRUCTION OF WELL-ADAPTED MODELS

We begin by proving a fundamental result:

Theorem 8.1. The functor $C^\infty(-): \mathrm{Mf} \longrightarrow \mathcal{B}^{op}$ preserves transversal pull-backs.

Since each of the full inclusions $FP\Pi_\infty \subseteq FG\Pi_\infty$ and $FG\Pi_\infty \subseteq \Pi_\infty\text{-Alg}$ preserve finite colimits, and $C^\infty(M) \in FP\Pi_\infty \subseteq \mathcal{B} \subseteq FG\Pi_\infty \subseteq \Pi_\infty\text{-Alg}$ it follows purely formally from the theorem that $C^\infty(-)$ preserves transversal pull-backs also when its value category is taken to be either $FP\Pi_\infty{}^{op}$, $FG\Pi_\infty{}^{op}$, or $\Pi_\infty\text{-Alg}^{op}$.

Proof. Several special cases have already been proved, e.g. Propositions 6.4 and 6.7. We prove (using these) two further special cases:

Proposition 8.2. Let $H \rightarrowtail M$ be a closed submanifold and $f: N \to M$ a smooth map, transversal to the inclusion $H \rightarrowtail M$. Then the (transversal) pull-back

$$\begin{array}{ccc} F & \longrightarrow & H \\ \downarrow & & \downarrow \\ N & \xrightarrow[f]{} & M \end{array}$$

($F = f^{-1}(H)$) is preserved by $C^\infty(-): \mathrm{Mf} \longrightarrow \mathcal{B}^{op}$.

Proposition 8.3. The functor $C^\infty(-): \mathrm{Mf} \longrightarrow \mathcal{B}^{op}$ preserves finite products.

Jointly, these two Propositions immediately lead to the theorem. This follows from the remarks following Definition 3.1 (a transversal pull-back of f_1, f_2 being represented as $(f_1 \times f_2)^{-1}(\Delta)$).

Proof of Proposition 8.2. We may cover M by open subsets $\{U_\alpha \mid \alpha \in A\}$ so that $H \cap U_\alpha$ is carved out of U_α by a function $h_\alpha: U_\alpha \longrightarrow \mathbb{R}^s$, as $h_\alpha^{-1}(\underline{0})$ with $\underline{0} \in \mathbb{R}^s$ a regular value of h_α. For each α, we have a diagram

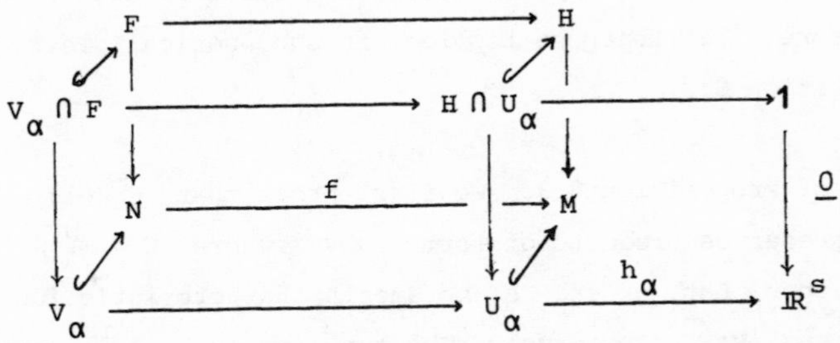

where $V_\alpha = f^{-1}(U_\alpha)$. All squares are (transversal) pull-backs in Mf, because $H \rightarrowtail M$ is transversal to f and $\underline{0}$ is a regular value for h_α. Proposition 6.4 applies to the right hand square in front, and to the composite of the two front squares. Purely diagrammatically, it follows that the left hand front square goes to a pull-back by $C^\infty(-)$. So our given square FHNM is 'covered' by pull-backs each of which is preserved by $C^\infty(-)$; we use sheaf theory to prove that the square FHNM itself goes to a pull-back, as follows. Identifying objects in Mf by their image in $\mathcal{B}^{op}$ under $C^\infty(-)$, let $X \in \mathcal{B}^{op}$, and let $n: X \to N$, $m: X \to H$ in $\mathcal{B}^{op}$ have $f \circ n = \text{incl} \circ m$, where incl is the inclusion of H into M. Let, for each α, X_α be formed by pulling $n: X \to N$ back along $V_\alpha \hookrightarrow N$. Since the V_α's cover N, the X_α's cover X, for the open cover topology in $\mathcal{B}^{op}$. From Proposition 6.7 it follows that the intersection $H \cap U_\alpha$ is also an intersection in $\mathcal{B}^{op}$, so that $X_\alpha \to X \to H$ factors across $H \cap U_\alpha$. Using that the left hand front square is a pull-back in $\mathcal{B}^{op}$, we get a map $\nu_\alpha: X_\alpha \to V_\alpha \cap F \subseteq F$, and having this for each $\alpha \in A$ gives an element $\{\nu_\alpha\}_{\alpha \in A}$ in

$$\prod_{\alpha \in A} \hom_{\mathcal{B}^{op}}(X_\alpha, F).$$

To see that the $\{\nu_\alpha\}$ form a compatible family, we use that $F \rightarrowtail N$ is monic in $\mathcal{B}^{op}$ (the restriction map $r: C^\infty(N) \longrightarrow C^\infty(F)$ being surjective, cf. the proof of Proposition 6.4). Using then the sheaf condition for the functor $\hom_{\mathcal{B}^{op}}(-, F)$ (Theorem 7.4) gives a global $X \to F$, composing with the $F \to N$ and $F \to H$ to the correct maps n and m, respectively. Uniqueness follows from the fact (Theorem 7.4) that the topology is subcanonical. This proves Proposition 8.2.

Proof of Proposition 8.3. We first prove that $C^\infty(-)$: $Mf \longrightarrow \mathcal{B}^{op}$ preserves products of form $U \times V$, where $U \subseteq \mathbb{R}^n$, $V \subseteq \mathbb{R}^m$ are open. Let g and h be smooth characteristic functions for U and V, respectively. The function $g \cdot h: \mathbb{R}^{n+m} \longrightarrow \mathbb{R}$ defined by

$$(g \cdot h)(\underline{x}, \underline{y}) := g(\underline{x}) \cdot h(\underline{y})$$

is then a smooth characteristic function for $U \times V \subseteq \mathbb{R}^{n+m}$. We have, using just the universal properties, that

$$C^\infty(\mathbb{R}^{n+m})\ \{(g \cdot h)^{-1}\} = C^\infty(\mathbb{R}^n)\ \{g^{-1}\} \otimes_\infty C^\infty(\mathbb{R}^m)\ \{h^{-1}\}\ .$$

But the three $\mathbb{T}_\infty$-algebras involved here are, by (6.7), just $C^\infty(U \times V)$, $C^\infty(U)$, and $C^\infty(V)$, respectively.

Next we prove that $C^\infty(-)$ preserves products of form $U \times M$, where $U \subseteq \mathbb{R}^n$, and M is a manifold. We do this by covering M with open sets V_α, where V_α is an open subset of $\mathbb{R}^m$. To pass from the preservation of the product $U \times V$ (already proved) to the preservation of $U \times M$, is then again a sheaf theoretic argument, like in the proof of Proposition 8.2. Finally, preservation of $N \times M$ is proved similarly, knowing the preservation of $U_\alpha \times M$ for a suitable open covering U_α of N.

We are now in a position to construct a well-adapted model in the sense of §§3 and 4. We shall in fact construct several later (§9 ff), but the one to be presented now is the most natural and comprehensive one.

We let

$$\mathcal{E} := \widetilde{\mathcal{B}^{op}} \subseteq \underline{\mathrm{Set}}^{\mathcal{B}} ,$$

the sheaves on $\mathcal{B}^{op}$ with respect to the open-cover topology of §7. We let $i\colon \mathrm{Mf} \longrightarrow \mathcal{E}$ be the composite

$$\mathrm{Mf} \xrightarrow{\;C^\infty(-)\;} \mathcal{B}^{op} \xrightarrow{\;y\;} \widetilde{\mathcal{B}^{op}} \tag{8.3}$$

where $y\colon \mathcal{B}^{op} \to \underline{\mathrm{Set}}^{\mathcal{B}}$ is the Yoneda embedding; it factors through the subcategory $\widetilde{\mathcal{B}^{op}}$ of sheaves, because the open cover topology on $\mathcal{B}^{op}$ is subcanonical (Theorem 7.4). We note that $i(\mathbb{R}) = y(C^\infty(\mathbb{R}))$ is just the forgetful functor $\mathcal{B} \longrightarrow \underline{\mathrm{Set}}$, since $C^\infty(\mathbb{R})$ is the free $\mathbb{T}_\infty$-algebra in one generator. With $i(\mathbb{R}) = R$, it is easy to prove

$$yB = \mathrm{Spec}_R(B)$$

for any $B \in \mathcal{B}$, in fact a more general equality holds, cf. the Note at the beginning of §9 below, notably formula (9.3).

<u>Theorem 8.4</u>. The model $i\colon \mathrm{Mf} \to \mathcal{E}$ described by (8.3) is a fully well-adapted model in the sense that the functor i is full and faithful and the Axioms A,B,C, and D of §§3 and 4 are satisfied.

<u>Proof</u>. The functor $C^\infty(-)$ is full and faithful by Corollary 5.8, and the Yoneda embedding y is full and faithful; hence

so is i. The functor $C^\infty(-)$ preserves transversal pull-backs (Theorem 8.1), and the Yoneda embedding preserves all limits that exist; hence i preserves transversal pull-backs. This is Axiom A. To prove Axiom B (= Axiom $1^W_{\mathbb{R}}$) for $R = i(\mathbb{R})$, we resort to the theory developed in §1; there, we talked about the algebraic theory $\mathbb{T}_k$ of k-algebras, but evidently this applies equally well to any algebraic theory $\mathbb{T}$ whatsoever, in particular to $\mathbb{T}_\infty$; $\otimes_k$ is to be replaced by $\otimes_\infty$, the coproduct in the category $\mathbb{T}_\infty$-Alg. We shall in particular use Theorem 1.2 for $\mathbb{T}_\infty$, with $\mathcal{R} = \mathcal{B}$:

We consider the canonical map (1.1)

$$\underset{\sim}{\mathrm{Hom}}_{R\text{-Alg}}(R^{\mathrm{Spec}_R(B)}, C) \xrightarrow{\nu_{B,C}} \mathrm{Spec}_C(B), \tag{8.4}$$

for $R = i(\mathbb{R})$ = forgetful functor $\mathcal{B} \longrightarrow \underline{\mathrm{Set}}$, $B \in FG\mathbb{T}_\infty$, and $C \in R\text{-Alg} = (R \downarrow \mathbb{T}_\infty\text{-Alg}(\mathcal{E}))$. From Theorem 1.2 follows that

Proposition 8.5. If $B \in \mathcal{B}$ is stable w.r.t $\mathcal{B}$ (meaning $B \otimes_\infty B' \in \mathcal{B}$ for any $B' \in \mathcal{B}$), then $\nu_{B,C}$ is an isomorphism for any $C \in R\text{-Alg}$.

Not all $B \in \mathcal{B}$ are stable in this sense: the coproducts in $\mathcal{B}$ are not in general preserved by $\mathcal{B} \hookrightarrow \mathbb{T}_\infty$-Alg, see Exercise 6.1. However, all Weil-algebras are:

Lemma 8.6. Let W be a Weil algebra over $\mathbb{R}$, and consider it with its canonical $\mathbb{T}_\infty$-structure. Then $B \in \mathcal{B} \Rightarrow B \otimes_\infty W \in \mathcal{B}$.

Proof. We have to see that $B \otimes_\infty W$ is germ-determined if B is. Since a Weil algebra W has a unique point $\pi: W \to \mathbb{R}$, it follows that the points $B \otimes_\infty W \longrightarrow \mathbb{R}$ of $B \otimes_\infty W$ are in 1-1-correspondence with those of B, via the projection $\bar{\pi}: B \otimes_\infty W \longrightarrow B$. By Theorem 5.3,

$$B \otimes_\infty W = B \otimes_{\mathbb{R}} W = B \oplus (B \otimes_{\mathbb{R}} W'),$$

with all elements in the second component nilpotent. Also, if $b = b_o + n \in B \oplus (B \otimes_{\mathbb{R}} W')$, then

$$(B \otimes_\infty W)\{b^{-1}\} = B\{b_o^{-1}\} \otimes_\infty W, \tag{8.5}$$

since n is nilpotent. From these observations easily follows that for any point $\bar{p}$ of B, and the corresponding point $p = \bar{p} \circ \bar{\pi}$ of $B \otimes_\infty W$, we have for the germ-algebras

$$(B \otimes_\infty W)_p \cong B_{\bar{p}} \otimes_\infty W . \tag{8.6}$$

To say that B is germ-determined is to say that the family

$$B \longrightarrow B_{\bar{p}} \;, \quad \bar{p} \text{ a point of } B$$

is jointly mono. The functor $-\otimes_{\mathbb{R}} W = -\otimes_\infty W$ preserves jointly monic families, since it, on the underlying vector spaces, is of form $C \longmapsto C \oplus \ldots \oplus C$ (m times, where m is the linear dimension of W over $\mathbb{R}$). It follows from (8.6), that the family $B \otimes_\infty W \longrightarrow (B \otimes_\infty W)_p$, p a point of $B \otimes_\infty W$, is jointly mono, so that $B \otimes_\infty W$ is germ-determined. This proves Lemma 8.6.

We conclude from Proposition 8.5 that the map $\nu_{B,C}$ in (8.4) is an isomorphism for $B = W$ any Weil-algebra over $\mathbb{R}$. On the other hand

$$\underset{\sim}{\mathrm{Hom}}_{(R\downarrow\amalg_{\mathbb{R}}\text{-}\mathrm{Alg}(E))}(R \otimes W, C) \cong \mathrm{Spec}_C(W) \tag{8.7}$$

by I.(16.3), and

$$\underset{\sim}{\mathrm{Hom}}_{(R\downarrow\amalg_{\mathbb{R}}\text{-}\mathrm{Alg}(E))}(R \otimes W, C) \cong \underset{\sim}{\mathrm{Hom}}_{(R\downarrow\amalg_{\infty}\text{-}\mathrm{Alg}(E))}(R \otimes W, C) \tag{8.8}$$

by an evident internalization of Theorem 5.3 (last clause). Combining the isomorphisms of (8.4), (8.7), and (8.8), we conclude by Yoneda's lemma, (as in I §16) that

$$\alpha\colon R \otimes W \longrightarrow R^{\mathrm{Spec}_R W}$$

is an isomorphism, proving Axiom B.

Validity of Axiom C is evident: $C^\infty(-)\colon \mathrm{Mf} \longrightarrow \mathcal{B}^{op}$ preserves open coverings, and $y\colon \mathcal{B}^{op} \longrightarrow \widetilde{\mathcal{B}^{op}}$ takes the coverings of the Grothendieck topology on $\mathcal{B}^{op}$ into jointly epic families in the topos it defines (this is a general fact about Grothendieck topologies).

We finally prove Axiom D. Since $\mathcal{B}$ has coproducts, we have from Proposition 1.4 that in $\underline{\mathrm{Set}}^{\mathcal{B}}$, every representable $y(B)$ is an atom. In particular, for B finitely presented, $y(B) = \mathrm{Spec}_R(B)$ is an atom in $\underline{\mathrm{Set}}^{\mathcal{B}}$. To prove that for $B = W$ a Weil-algebra, $yW = \mathrm{Spec}_R W$ is an atom in $\widetilde{\mathcal{B}^{op}}$, it suffices to see that $F_{yW} \in \underline{\mathrm{Set}}^{\mathcal{B}}$ is a sheaf if F is, where $(-)_{yW}$ denotes the right adjoint of $(-)^{yW}$. This follows purely formally if we can prove

$$(aX)^{yW} = a(X^{yW})$$

where X is an arbitrary object in $\underline{\mathrm{Set}}^{\mathcal{B}}$, and $a\colon \underline{\mathrm{Set}}^{\mathcal{B}} \to \widetilde{\mathcal{B}^{op}}$ is the sheaf reflection functor. (For then, for $F \in \widetilde{\mathcal{B}^{op}}$ and $X \in \underline{\mathrm{Set}}^{\mathcal{B}}$, we have, (denoting hom-sets by square brackets)

$$[aX, F_{yW}] = [(aX)^{yW}, F] = [a(X^{yW}), F]$$

$$= [X^{yW}, F] = [X, F_{yW}].\ )$$

To prove $a(X^{yW}) = (aX)^{yW}$, we use the classical construction of a as $\ell \circ \ell$ ([1], II.3) where $\ell\colon \underline{\mathrm{Set}}^{\mathcal{B}} \longrightarrow \underline{\mathrm{Set}}^{\mathcal{B}}$ is the functor

$$\ell(X)(B) := \varinjlim_{R \hookrightarrow yB} \hom(R, X)$$

the colimit ranging over 'covering cribles' $R \hookrightarrow yB$. Then

$$\ell(X^{yW})(B) = \varinjlim_{R \hookrightarrow yB} \hom(R, X^{yW}) = \varinjlim_{R \hookrightarrow yB} (R \times yW, X) \tag{8.9}$$

whereas

$$(\ell\ X)^{yW}(B) = (\ell X)(W \otimes B) = \varinjlim_{R' \hookrightarrow y(W \otimes B)} \hom(R',X). \qquad (8.10)$$

These two objects are isomorphic: for, since every open (co-) cover of $W \otimes_\infty B$ in $\mathcal{B}$ comes about from a unique open (co-) cover of B, by pushing out along $B \to W \otimes_\infty B$ (by an argumentation as in the proof of Lemma 8.6), it follows that the index categories used for the colimits in (8.9) and (8.10) are equivalent, and if $R' \hookrightarrow y(W \otimes B)$ corresponds to $R \hookrightarrow yB$, $R' = R \times yW$. From $\ell(X^{yW}) \cong (\ell X)^{yW}$ follows $a(X^{yW}) \cong (aX)^{yW}$, as desired. This proves Axiom D.

EXERCISES

8.1 (Grothendieck-Verdier,[1]). Define a Grothendieck pre-topology on Mf by declaring a family $\{M_i \to M \mid i \in I\}$ to be a covering if it is jointly surjective and each $M_i \to M$ is an open inclusion. Prove that it is subcanonical. Let $\widetilde{Mf}$ be the resulting topos ("the smooth topos"). Prove that the functor $y\colon Mf \longrightarrow \widetilde{Mf}$ is a well-adapted model in the sense that Axioms A, C, and D of §§3 and 4 are satisfies. (Hint: Axiom D is satisfied for the trivial reason that $Spec_R(W) = \mathbf{1}$ for any Weil algebra W, so no proper infinitesimal objects exist, ("R has no nilpotents").) Prove that Axiom B fails.

However (Reyes), synthetic calculus in Fermat-style (I,Axiom (2.4)) is available, cf. Exercise 9.4 below.

III.9: W-DETERMINED ALGEBRAS, AND MANIFOLDS WITH BOUNDARY

To get toposes in which the category of manifolds with boundary is nicely and fully embedded, it seems necessary to construct some "smaller" well-adapted models, by choosing suitable full subcategories $\mathcal{C} \subseteq \mathcal{B}$ as our site of definition. In order to be able to utilize the open-cover-topology (§7), we shall assume that $\mathcal{C}$ satisfies the following condition:

If $B \in \mathcal{C}$ and $B \rightarrow C$ is an open inclusion in $\mathcal{B}$, then $C \in \mathcal{C}$ (9.1)

(recalling Definition 7.1 of the notion 'open inclusion'). Since the pushout of an open inclusion along any map is an open inclusion, and the composite of two open inclusions is an open inclusion, it is clear that the class of open co-coverings of objects of $\mathcal{C}$ form a Grothendieck pre-co-topology on $\mathcal{C}$ (likewise called the open cover topology), and that any sheaf $F: \mathcal{B} \longrightarrow \underline{\text{Set}}$ has the property that its restriction to $\mathcal{C}$ is a sheaf. In particular, the topology on $\mathcal{C}$ is subcanonical. If

$$j: \mathcal{C} \hookrightarrow FG\amalg_\infty$$

denotes the inclusion functor, some specially useful sheaves are functors of form

$$\hom_{FG\amalg_\infty}(B, j(-)): \mathcal{C} \longrightarrow \underline{\text{Set}} \tag{9.2}$$

for $B \in FG\amalg_\infty$. If B is not in $\mathcal{C}$, such a functor is not representable. We say nevertheless that B <u>represents it "from the outside"</u>, and write $y_o(B)$ for the functor (9.2).

Examples. 1) Let $\mathcal{C}_S \subseteq \mathcal{B}$ be the full image of the functor $C^\infty(-): Mf^{op} \longrightarrow \mathcal{B}$. It satisfies condition (9.1); this $\mathcal{C}_S$ leads to the smooth topos (cf. Exercise 8.1).

2) Let $\mathcal{C}_C \subseteq \mathcal{B}$ be the category of $\amalg_\infty$-algebras of form

$$C^\infty(M) \otimes_\infty W \quad (= C^\infty(M) \otimes_{\mathbb{R}} W)$$

for W a Weil algebra over $\mathbb{R}$, and $M \in Mf$. The fact that (9.1) holds follows from (8.5). This $\mathcal{C}_C$ leads to the "Cahiers topos" of Dubuc [11].

3) $\mathcal{C} = \mathcal{B}$. This leads to the topos considered in §8.

4) $\mathcal{C} = \mathcal{W}$, the category of Weil algebras. This rather trivial example is studied in Exercise 9.9.

Note. The functor

$$y_0: (FG\amalg_\infty)^{op} \longrightarrow \widetilde{\mathcal{C}}^{op}$$

is proper left exact and factors through the reflection functor

$$(FG\amalg_\infty)^{op} \xrightarrow{\ \wedge\ } \mathcal{B}^{op}$$

(which is also proper left exact, being a right adjoint). For, since $\mathcal{C} \subseteq \mathcal{B}$,

$$hom_{\amalg_\infty\text{-}Alg}(B, j(-)) \cong hom_{\amalg_\infty\text{-}Alg}(\hat{B}, j(-)),$$

where $j: \mathcal{C} \longrightarrow FG\amalg_\infty$ is the inclusion functor.

Note also that since y_0 is proper left exact, and

$$y_0(C^\infty(\mathbb{R})) = R = \text{forgetful functor } \mathcal{C} \longrightarrow \underline{Set}$$

$(= y(C^\infty(\mathbb{R}))$ if $C^\infty(\mathbb{R}) \in \mathcal{C})$, we have for any $B \in FG\amalg_\infty$

$$Spec_R(B) = y_0(B) = y_0(\hat{B}) \tag{9.3}$$

The examples $\mathcal{C}_S, \mathcal{C}_C$, and $\mathcal{W}$, for $\mathcal{C}$, have the property that all their objects are W-determined $\mathbb{T}_\infty$-algebras in the sense of the following Definition 9.1. This notion is analogous to, but stronger than, the notion germ-determined in §6.

Definition 9.1. Let $B \in \mathbb{T}_\infty$-Alg, and let $I \subseteq B$ be an ideal. The W-radical (W for "Weil"), denoted $\bar{I}$, is the ideal in B given by

$$\bar{I} := \{f \in B \mid \varphi(f) = 0 \quad \text{for any } \mathbb{T}_\infty\text{-algebra map (or } \mathbb{T}_{\mathbb{R}}\text{-algebra map, cf. Exercise 5.1) } \varphi\colon B \to W \text{ with } \varphi(I) = 0 \text{ into any Weil algebra } W\}.$$

We say that B is W-determined if $\overline{\{0\}} = \{0\}$.

It is possible to prove that, for $B = C^\infty(\mathbb{R}^n)$ and $I \subseteq B$ an ideal, the W-radical is exactly the closure of I in the Whitney topology on $C^\infty(\mathbb{R}^n)$, cf. [69]. We shall not utilize this.

Proposition 9.2. Each $\mathbb{T}_\infty$-algebra of form $C^\infty(M) \otimes_\infty W$ is W-determined.

Proof. The family $\{ev_p\colon C^\infty(M) \to \mathbb{R} \mid p \in M\}$ is jointly mono. Hence so is the family

$$\{ev_p \otimes id\colon C^\infty(M) \otimes_\infty W \longrightarrow W \mid p \in M\}$$

since $C^\infty(M) \otimes_\infty W = C^\infty(M) \otimes_{\mathbb{R}} W$, and the functor $-\otimes_{\mathbb{R}} W$ preserves jointly monic families (as we observed in the proof of Proposition 8.6).

Note that since a Weil algebra W has a unique point $W \to \mathbb{R}$, every $q\colon B \to W$ $(B \in \mathbb{T}_\infty\text{-Alg})$ defines a point p of B, namely the composite $B \to W \to \mathbb{R}$. We say that $q\colon B \to W$ is centered at the point p. If $f \in B$ has $p(f)$ invertible (i.e. $\neq 0$) in $\mathbb{R}$, $q\colon B \to W$ factors across $B\{f^{-1}\}$ (recall that $W \to \mathbb{R}$ reflects the property of being invertible, cf. the proof of Proposition I.19.1). Since this holds for all f with $p(f)$ invertible, we see that any $B \to W$, centered at p, factors across $B \to B_p = B\{\Sigma_p^{-1}\}$, where $\Sigma_p = \{f \in B \mid p(f)$ is invertible$\}$. It follows in particular that a W-determined algebra is also germ-determined. A point-determined Π_∞-algebra is clearly W-determined.

The category of W-determined algebras is a reflective subcategory of the category of all Π_∞-algebras: to any Π_∞-algebra B, we may divide out by the W-radical of the zero ideal to get a W-determined algebra.

We now begin the consideration of manifolds-with-boundary. The fundamental such is the non-negative half-line

$$\mathbb{H} := \{x \in \mathbb{R} \mid x \geq 0\}$$

Then $C^\infty(\mathbb{H}) = C^\infty(\mathbb{R})/I_1$ where I_1 is the ideal of functions vanishing on $\mathbb{H}$.

It is clear that if $f \in I_1$, then $f^{(k)}(0) = 0 \;\forall k$, i.e. f is <u>flat</u> at 0. So any $f \in I_1$ can by Hadamard's lemma, be written

$$f(x) = x^{k+1} \cdot g(x)$$

for any $k = 1,2,\ldots$, in particular $I \subseteq (x^{k+1})$. So we have a surjective Π_∞-homomorphism

$$C^\infty(\mathbb{R})/I_1 = C^\infty(\mathbb{H}) \twoheadrightarrow C^\infty(\mathbb{R})/(x^{k+1}) = \mathbb{R}[X]/(X^{k+1}) \qquad (9.4)$$

(the last equality by Proposition 5.11).

We consider in $C^\infty(\mathbb{R}^{n+1}) = C^\infty(\mathbb{R}^n) \otimes_\infty C^\infty(\mathbb{R})$ the ideal I generated by the image of I_1 under the inclusion of $C^\infty(\mathbb{R})$ into this tensor product. So I consists of functions of form

$\sum g_i(\underline{x},y) \cdot h_i(y)$

(finite sum) with $h_i \in I_1$.

<u>Proposition 9.3.</u>* The W-radical $\bar{I}$ of I consists of those functions $f: \mathbb{R}^n \times \mathbb{R} \to \mathbb{R}$ which vanish on $\mathbb{R}^n \times \mathbb{H}$.

<u>Proof</u>. Let $\varphi: C^\infty(\mathbb{R}^n \times \mathbb{R}) \longrightarrow W$ annihilate the ideal I, and let $f: \mathbb{R}^n \times \mathbb{R} \longrightarrow \mathbb{R}$ vanish on $\mathbb{R}^n \times \mathbb{H}$. We shall prove $\varphi(f) = 0$, which will prove $f \in \bar{I}$. Clearly, φ is centered at some point $p = (\underline{b},a) \in \mathbb{R}^n \times \mathbb{H}$. By Proposition 5.12, it is of form (identifying W with $\mathbb{R}^k$)

$$P_j\left(\frac{\partial}{\partial x_1}, \ldots, \frac{\partial}{\partial x_n}, \frac{\partial}{\partial y}\right)(\ )(p) \quad j = 1,\ldots,k$$

for a k-tuple of polynomials P_j in n+1 variables, all of degree $\leq K$, say. We can, by Hadamard's lemma, write the given f in form

$$f(\underline{x},y) = \sum_{|\alpha| \leq K} \frac{1}{\alpha!}(\underline{x}-\underline{b})^\alpha \cdot \frac{\partial^\alpha f}{\partial x^\alpha}(\underline{b},y) + \sum_{|\beta|=K+1} (\underline{x}-\underline{b})^\beta \cdot h_\beta(\underline{x},y)$$

where α and β are multi-indices in $(1,\ldots,n)$. Since f vanishes on $\mathbb{R}^n \times \mathbb{H}$, we have

$$\frac{\partial^\alpha f}{\partial x^\alpha}(\underline{b},y) \in I_1,$$

and likewise $h_\beta(\underline{x},y)$ vanishes on $\mathbb{R}^n \times \mathbb{H}$. So in the above sum, the α-sum belongs to I and is thus killed by φ. So it suffices to see that the β-summation is killed by <u>any</u> differential operator of form

$$P\left(\frac{\partial}{\partial x_1}, \ldots, \frac{\partial}{\partial x_n}, \frac{\partial}{\partial y}\right)(-)(\underline{b},a),$$

---

*Quê and Reyes have announced a stronger result which implies that I itself contains all functions vanishing on $\mathbb{R}^n \times \mathbb{H}$.

with P a polynomial of degree $\leq K$. We write P in form $\Sigma Q_\ell(X_1,\ldots,X_n)\cdot Y^\ell$, with the Q_ℓ's of degree $\leq K$. So

$$P(\frac{\partial}{\partial x_1},\ldots,\frac{\partial}{\partial x_n},\frac{\partial}{\partial y})((\underline{x}-\underline{b})^\beta\cdot h_\beta(\underline{x},y))(\underline{b},a)$$

$$= \sum_\ell Q_\ell(\frac{\partial}{\partial x_1},\ldots,\frac{\partial}{\partial x_n})((\underline{x}-\underline{b})^\beta\cdot\frac{\partial^\ell}{\partial y^\ell}\ h_\beta(\underline{x},y))(\underline{b},a).$$

Now, none of the differential operators $Q_\ell(\frac{\partial}{\partial x_1},\ldots,\frac{\partial}{\partial x_n})$ have high enough degree to derive away all of the factors $(\underline{x}-\underline{b})^\beta$ because $|\beta| = K+1$. So when substituting $\underline{b}$ for $\underline{x}$ after the differentiation, we get 0.

In the rest of this § we shall make the following set of assumptions on $\mathcal{C}$:

<u>Assumption</u> (i) $\mathcal{C} \subseteq \mathcal{B}$, and $\mathcal{C}$ satisfies (9.1). (ii) $C^\infty(M)$ is in $\mathcal{C}$ for any $M \in \mathrm{Mf}$; (iii) if $C \in \mathcal{C}$, then so is $C \otimes_\infty W$, for any Weil algebra W; (iv) each $C \in \mathcal{C}$ is W-determined.

By Proposition 9.2, an example of such $\mathcal{C}$ is $\mathcal{C}_C$, the class of $\mathbb{T}_\infty$-algebras of form $C^\infty(M) \otimes_\infty W$.

Under the Assumption, we have a functor, analogous to (8.3), namely

$$\mathrm{Mf} \xrightarrow{\ C^\infty(-)\ } \mathcal{C}^{op} \xrightarrow{\ y\ } \widetilde{\mathcal{C}^{op}} \tag{9.5}$$

where $y: \mathcal{C}^{op} \to \mathrm{Set}^{\mathcal{C}}$ is the Yoneda embedding; it factors through the subcategory $\widetilde{\mathcal{C}^{op}}$ of sheaves, because the open-cover topology on $\mathcal{C}^{op}$ is subcanonical. We note that $i(\mathbb{R}) = y(C^\infty(\mathbb{R}))$ is just the forgetful functor $\mathcal{C} \longrightarrow \underline{\mathrm{Set}}$, since $C^\infty(\mathbb{R})$ is the free $\mathbb{T}_\infty$-algebra in one generator.

Under the Assumption above, we have, writing $\mathcal{E}$ for $\widetilde{\mathcal{C}^{op}}$:

<u>Theorem 9.4</u>. The model $i: \mathrm{Mf} \to \mathcal{E}$ thus described is a fully well-adapted model in the sense that the functor i is full and faithful, and the Axioms A,B,C, and D of §§ 3 and 4 are satisfied.

Proof. This is a repetition of the proof of Theorem 8.4, with $\mathcal{C}$ instead of $\mathcal{B}$. Note that $\mathcal{C}^{op} \hookrightarrow \mathcal{B}^{op}$ does not preserve finite limits in general, but since $C^\infty(-): \text{Mf} \longrightarrow \mathcal{B}^{op}$ preserves transversal pull-backs and factors through $\mathcal{C}^{op} \subseteq \mathcal{B}^{op}$, $C^\infty(-): \text{Mf} \rightarrow \mathcal{C}^{op}$ does preserve transversal pull-backs. Lemma 8.6 is replaced by (iii) in the Assumption.

The rest of the § is concerned with extending $i: \text{Mf} \longrightarrow \mathcal{E}$ to a full and faithful functor on Mf', the category of manifolds-with-boundary, and investigating properties of this extension.

We do this as follows. If K is a manifold-with-boundary, one may find a manifold M without boundary, of same dimension, containing K, and such that each $x \in \partial K$ (= boundary of K) has an open neighbourhood U in M diffeomorphic to $\mathbb{R}^n$, by a diffeomorphism which takes $U \cap K$ into $\mathbb{R}^{n-1} \times \mathbb{H} \subseteq \mathbb{R}^n$. We then have

$$C^\infty(K) \cong C^\infty(M)/I$$

where I is the ideal of functions $f: M \rightarrow \mathbb{R}$ vanishing on K. It is clear that I is germ-determined. Since $C^\infty(M) \in \mathcal{B}$ (i.e. finitely generated and germ-determined), the same holds true for $C^\infty(K)$. From Theorem 5.7, it follows that a point of $C^\infty(K)$ must be evaluation at some $P \in M$, and it is clear that this P cannot be outside $K \subseteq M$ (which is a closed subset). So the notational and terminological confusion between points of K and points of $C^\infty(K)$ is justified, just as for manifolds-without-boundary.

In particular, each $C^\infty(K)$ is point-determined, and hence also W-determined.

If Mf' denotes the category of manifolds-with-boundary (it contains Mf as a full subcategory), we extend the functor $i: \text{Mf} \longrightarrow \mathcal{E} = \widetilde{\mathcal{C}}^{op}$ into a functor $i': \text{Mf}' \longrightarrow \mathcal{E}$ by letting $i'(K)$ be the functor represented-from-the-outside by $C^\infty(K)$,

$$i'(K) = \hom_{\mathbb{T}_\infty\text{-Alg}}(C^\infty(K), j(-)): \quad \mathcal{C} \longrightarrow \underline{\text{Set}},$$

which is a sheaf, as we noted at the beginning of the §. It is now easy to prove

<u>Theorem 9.5</u>. The functor $i': Mf' \longrightarrow \mathcal{E} = \mathcal{C}^{op}$ preserves any product of form $N \times K$, where $N \in Mf \subseteq Mf'$ and $K \in Mf'$.

<u>Proof</u>. Let $K \subseteq M$ be as above, with $M \in Mf$, a manifold-without-boundary. Then

$$C^{\infty}(N \times K) = C^{\infty}(N \times M)/J$$

where J is the ideal of functions vanishing on $N \times K \subseteq N \times M$. On the other hand

$$C^{\infty}(N) \otimes_{\infty} C^{\infty}(K) = C^{\infty}(N) \otimes_{\infty} (C^{\infty}(M)/I_1) = (C^{\infty}(N) \otimes_{\infty} C^{\infty}(M))/I,$$

where I_1 is the ideal in $C(M)$ of functions vanishing on K, and I is the ideal which I_1 generates in $C^{\infty}(N) \otimes_{\infty} C^{\infty}(M)$. We claim* that the W-radical $\bar{I}$ of I is J. Clearly $I \subseteq J$, and since J is W-determined, $\bar{I} \subseteq J$. Conversely, let $f \in J$. To prove $f \in \bar{I}$, it suffices, since $\bar{I}$ is W-determined, hence germ-determined, to prove $f_p \in (\bar{I})_p$ for all points p of $N \times M$. But now that we are working with germs, we may as well assume that $N = \mathbb{R}^n$, $M = \mathbb{R}^m$, $K = \mathbb{R}^{m-1} \times \mathbb{H}$; then J consists of functions vanishing on $\mathbb{R}^n \times \mathbb{R}^{m-1} \times \mathbb{H} \subseteq \mathbb{R}^{n+m}$, and I contains any $h(x_1, \ldots, x_n, y_1, \ldots, y_m)$ which only depends on y_m and vanishes if $y_m \geq 0$. Then $f \in \bar{I}$ follows from Proposition 9.3.

It follows that for any W-determined $\mathbb{T}_{\infty}$-algebra C,

$$\hom_{\mathbb{T}_{\infty}}(C^{\infty}(N \times M)/J, C) \cong \hom_{\mathbb{T}_{\infty}}(C^{\infty}(N) \otimes_{\infty} C^{\infty}(K), C) ;$$

since $\mathcal{C}$ is assumed to consists of W-determined algebras, it

---

* the result announced by Quê and Reyes (mentioned earlier) implies that in fact $I = J$ $(= \bar{I})$.

follows that $C^\infty(N \times M)/J$, i.e. $C^\infty(N \times K)$, and $C^\infty(N) \otimes_\infty C^\infty(K)$ represent (from the outside) the same object in $\widetilde{C^{op}}$. The Theorem now follows.

The last endeavour of this § is to prove

Theorem 9.6. The functor $i': Mf' \longrightarrow E = \widetilde{C^{op}}$ is full and faithful.

Proof. We have the functor $\Gamma = \hom(\mathbf{1},-): E \to \underline{Set}$. If $K \in Mf'$

$$\Gamma(i'K) = \hom_E(y(\mathbb{R}), y_o(C^\infty(K))) = \hom_{\mathbb{T}_\infty\text{-Alg}}(C^\infty(K), \mathbb{R}) = |K|, \tag{9.6}$$

the underlying set of K. Since the "underlying-set" functor $|\cdot|: Mf' \longrightarrow \underline{Set}$ is faithful, it follows that i' is. To prove it full, let $f: i'K_1 \longrightarrow i'K_2$ be a map in E. Applying Γ, we get by (9.6) a map $\Gamma(f)$ in $\underline{Set}$ from $|K_1|$ to $|K_2|$. We claim that it is smooth. First, by considering a boundaryless manifold $M_2 \supseteq K_2$, as above, we see that we may replace K_2 by M_2, or changing notation, assume that K_2 itself is boundaryless, $K_2 \in Mf$. For any $N \in Mf$ and any smooth $g: N \to K_1$, we have

$$f \circ i(g): iN \longrightarrow iK_2,$$

and since i is full, this comes from a smooth map $N \to K_2$, whose underlying point-set map $|N| \to |K_2|$ is $\Gamma(f) \circ |g|$. So we have proved that $\Gamma(f): |K_1| \longrightarrow |K_2|$ is plot smooth, where we pose

Definition 9.7. A point-set map $h: |K_1| \to |K_2|$ where $K_1 \in Mf'$, $K_2 \in Mf$ is plot smooth if its composite with any smooth $g: N \to K_1$ (with $N \in Mf$) is smooth.

The g's occurring here are to be thought of as "test plots", whence the name "plot smooth". The concept may be applied to any subset K_1 of a manifold.

The main lemma in the proof is the following, whose proof depends on a well known theorem of Whitney concerning smooth even functions [80]:

Lemma 9.8. Any plot smooth $h: |K_1| \to |K_2|$ (with K_1 and $K_2 \in Mf'$) is smooth.

Proof. To prove h smooth, it suffices to prove its composite with any smooth $K_2 \to \mathbb{R}$ to be smooth. Since the composite of a plot smooth map with a smooth map clearly is plot smooth, it suffices to prove the Lemma for the case $K_2 = \mathbb{R}$.

Let $K_1 \subseteq M$ with M boundaryless. To prove h smooth at an interior point p of K_1 is obvious. So consider the case $p \in \partial K_1$, the boundary of K_1. To say h smooth at p means that it can be extended to a smooth function on some open $U \subseteq M$ containing p. Since the question whether this can be done is local, we may as well assume that $K_1 = \mathbb{R}^{n-1} \times \mathbb{H}$, $M = \mathbb{R}^n$, and $p = (\underline{b}, 0)$ ($\underline{b} \in \mathbb{R}^{n-1}$). We consider the plot s

$$\mathbb{R}^n \xrightarrow{\ s\ } \mathbb{R}^{n-1} \times \mathbb{H} \tag{9.7}$$

given by

$$(\underline{x}, y) \longmapsto (\underline{x}, y^2).$$

By assumption of plot smoothness of h, $h \circ s$ is smooth $\mathbb{R}^n \to \mathbb{R}$, and it clearly is even in the last variable y (i.e. takes same value on $(\underline{x}, y)$ and $(\underline{x}, -y)$). By Whitney's theorem quoted above, there exists a smooth $f: \mathbb{R}^n \longrightarrow \mathbb{R}$ such that

$$f(\underline{x}, y^2) = h(s(\underline{x}, y)) \quad (= h(\underline{x}, y^2)) \quad \forall (\underline{x}, y) \in \mathbb{R}^n .$$

Since every $z \in \mathbb{H}$ is of form y^2 for some y, it follows that

f and h agree on $\mathbb{R}^{n-1} \times \mathbb{H}$. So h has been extended to a smooth map (namely f) defined on all of $\mathbb{R}^n$. This proves the Lemma.

Returning to the proof of the theorem, we have for $f: i'K_1 \to i'K_2$ that $\Gamma f: |K_1| \to |K_2|$ is plot smooth, hence smooth, by the Lemma; so we have

$$i'\Gamma f: \; i'K_1 \longrightarrow i'K_2.$$

If we can prove $i'\Gamma f = f$ we have proved i' full. Clearly $\Gamma(i'\Gamma f) = \Gamma(f)$. We may embed K_2 into a boundaryless manifold M_2, as above, and then, by Whitney's embedding theorem, embed M_2 as a closed submanifold of some $\mathbb{R}^m$. These two embeddings go by i' to monic maps. To see $i'\Gamma f = f$, we may therefore reduce to the case where $K_2 = \mathbb{R}^m$, and then, since i preserves products, to the case $K_2 = \mathbb{R}$. The result then follows (by taking $K = K_1$ and $g = f - i'\Gamma f$) from

Lemma 9.9. Let $K \in Mf'$, and let $g: i'K \to R$ in $\mathcal{E} = \widetilde{\mathcal{C}}^{op}$ have the property that $\Gamma(g): K \to \mathbb{R}$ is the zero map. Then g itself is the zero map.

Proof. It suffices to see that for any $yC \to i'K$ with $C \in \mathcal{C}$, the composite $yC \longrightarrow i'K \to R$ is zero. Since $R = y(C^\infty(\mathbb{R}))$, this composite is induced by a $\mathbb{T}_\infty$-homomorphism $\gamma: C^\infty(\mathbb{R}) \longrightarrow C$, which we have to see sends $id_{\mathbb{R}}$ into the zero element of C. Since C is W-determined, it suffices to see that each composite

$$C^\infty(\mathbb{R}) \xrightarrow{\;\gamma\;} C \longrightarrow W,$$

with W a Weil algebra, sends $id_{\mathbb{R}}$ to 0. This reduces the problem to proving that for any $b: yW \longrightarrow i'K$ (with W a Weil algebra), the composite

$$yW \xrightarrow{\;b\;} i'K \xrightarrow{\;g\;} R$$

is zero. Since $i'K = y_o(C^\infty(K))$, b is induced by a $\mathbb{T}_\infty$-homomorphism $C^\infty(K) \to W$, centered at some $p \in K$, and therefore factoring through $C^\infty(K) \to C^\infty_p(K)$. We then have the situation

$$\begin{array}{ccccc}
yW & \xrightarrow{\;b\;} & y_o(C^\infty(K)) & \xrightarrow{\;g\;} & y(C^\infty(\mathbb{R})) \\
 & \searrow^{\bar{b}} & \uparrow & \nearrow_{\bar{g}} & \\
 & & y_o(C_p(K)) & &
\end{array}$$

and it suffices to prove $\bar{g} \circ \bar{b}$ to be the zero map. Since we are now working with germs, we may as well assume that $K = \mathbb{R}^{n-1} \times \mathbb{H}$, and $p = (\underline{x},0)$ (the case $p = (\underline{x},y)$ with $y > 0$ is quickly reduced to the boundaryless case). Now it is clear that b factors through

$$R^{n-1} \times D_k \subseteq R^{n-1} \times i'(\mathbb{H})$$

for some k (the inclusion $D_k \subseteq i'(\mathbb{H})$ is by (9.4)). So it suffices to see that our $g\colon R^{n-1} \times i'(\mathbb{H}) \to R$ vanishes on $R^{n-1} \times D_k$. We fire the final shot by once again pulling the "squaring map" (cf. (9.7)); we have, writing H for $i'(\mathbb{H})$,

$$R^n \xrightarrow{\;i(s)\;} R^{n-1} \times H \longrightarrow R \tag{9.8}$$

Since $i\colon \mathrm{Mf} \to \mathcal{E}$ is full, this composite is of form $i(\ell)$ for some smooth $\ell\colon R^n \to R$, and since g is zero on global points $p\colon \mathbf{1} \to R^{n-1} \times H$, i.e. on elements of $\mathbb{R}^{n-1} \times \mathbb{H}$, it follows that ℓ must be the zero map, whence also (9.8) is the zero map. Therefore also

$$R^{n-1} \times D_{2k+1} \xrightarrow{\;id \times s\;} R^{n-1} \times D_k \xrightarrow{\;g\;} R$$

is the zero map, where $s\colon D_{2k+1} \longrightarrow D_k$ is the restriction of the squaring map. Taking exponential adjoints, we get that

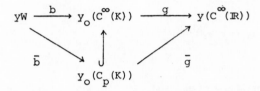

$$R^{n-1} \xrightarrow{\hat{g}} R^{D_k} \xrightarrow{R^s} R^{D_{2k+1}}$$

is the zero map. But "R believes $s: D_{2k+1} \longrightarrow D_k$ is epic", meaning $R^s: D^{D_k} \rightarrow R^{D_{2k+1}}$ is monic; this follows from Axiom 1^W and the fact that we have a monic map between the Weil algebras for D_k and D_{2k+1},

$$\mathbb{R}[X]/(X^{k+1}) \longrightarrow \mathbb{R}[Y]/(Y^{2k+2})$$

given by $X \mapsto Y^2$.

Since $R^s \circ \hat{g}$ is the zero map, and R^s is monic, $\hat{g}$ is the zero map, hence $g: R^{n-1} \times D_k \rightarrow R$ is the zero map. Lemma 9.9 is proved. So Theorem 9.6. is proved.

EXERCISES

9.1. Define a functor $i': Mf' \longrightarrow \widetilde{\mathcal{B}}^{op}$ by

$$i'(K) := y(C^\infty(K)).$$

Prove that i' is full and faithful. (Hint: we have remarked that Theorem 5.7 is valid for K as well. Now copy the proof of Corollary 5.8.)

9.2. Use the result of Quê and Reyes quoted in the footnotes to prove that the functor i' considered in Exercise 9.1 preserves products of form $N \times K$ with $N \in Mf$ and $K \in Mf'$.

9.3. Let $I' \subseteq C^\infty(\mathbb{R})$ be the ideal of functions g for which there exists an open $U \supseteq \mathbb{H}$ on which g vanishes. Prove that the W-radical of I' is the ideal of functions vanishing on $\mathbb{H}$. Prove that it is strictly larger than I'. However, the germ radical of I' is I' itself. (Hint: use Proposition 5.12.)

9.4 (Reyes). Let $\mathcal{C} \subseteq \mathcal{B}$ satisfy (9.1), and assume $C^\infty(\mathbb{R}^m) \in \mathcal{C}$ $\forall m = 0,1,\ldots.$ Finally assume $\forall B \in \mathcal{C}: C^\infty(\mathbb{R}^m) \otimes_\infty B \in \mathcal{C}$

$\forall$ m. If $I \subseteq C^{\infty}(\mathbb{R}^m)$ is an ideal, let $I^* \subseteq C^{\infty}(\mathbb{R}^{m+1})$ be the ideal generated by I under the inclusion $C^{\infty}(\mathbb{R}^m) \longrightarrow C^{\infty}(\mathbb{R}^m) \otimes_{\infty} C^{\infty}(\mathbb{R}) = C^{\infty}(\mathbb{R}^{m+1})$. Consider a generalized element $f \in_{yB} R^R$, where $B = C^{\infty}(\mathbb{R}^m)/I$. Prove that the information contained in f is given by an $F \in C^{\infty}(\mathbb{R}^{m+1})$ modulo I^*. Use this, in connection with Exercise 5.1, to prove that

$$\vdash_1 \forall f \in R^R (f \cdot \mathrm{id} \equiv 0 \Rightarrow f \equiv 0).$$

Similarly, using a parametrized form of Hadamard's lemma, prove that

$$\vdash_1 \forall f \in R^R \; \exists g \in R^{R\times R}: \; \forall x,y \;\; f(x) - f(y) = (x-y)\cdot g(x,y)$$

and deduce (cf. Exercise I.13.2) that Fermat's axiom I.(2.4) holds, thus some synthetic calculus is available.

In particular, prove that Fermat's Axiom holds for the Smooth Topos $\widetilde{\mathcal{C}_S^{op}}$.

Reyes and others have observed that in this model, the object R_{ded} of Dedekind reals cf. [28], 6.61) is represented from the outside by $C^0(\mathbb{R})$ (= ring of continuous functions on $\mathbb{R}$). The inclusion $C^{\infty}(\mathbb{R}) \subseteqq C^0(\mathbb{R})$ gives rise to a comparison map $R \to R_{ded}$ which has been utilized and studied by Reyes and Veit.

9.5. Let $\chi: \mathbb{R} \longrightarrow \mathbb{R}$ be a smooth characteristic function for $\mathbb{N} = \{x \in \mathbb{R} \mid x < 0\}$, i.e. $\chi(x) \neq 0$ iff $x < 0$. The ideal (χ) is germ-determined by Theorem 6.3. Prove that its W-radical consists of functions vanishing on $\mathbb{H}$. (Hint: use Proposition 5.12.)

9.6. We may consider the following on a $\mathbb{T}_{\infty}$-algebra object R

<u>Axiom B.2</u>. For any $B \in \mathcal{C}_C$ (i.e. $B = C^{\infty}(M) \otimes_{\infty} W$),

$$\underset{\sim\sim}{\mathrm{Hom}}_{R\text{-Alg}}(R^{\mathrm{Spec}_R B}, C) \xrightarrow{\;\cong\;} \mathrm{Spec}_C(B)$$

is an isomorphism.

Adapt the proof of Theorem 1.2 to prove that Axiom B.2 holds for $\widetilde{\mathcal{C}}^{op}_{C}$. (Hint: the inclusion functor $\mathcal{C} \subseteq \Pi_\infty$-Alg preserves finite coproducts, by an adaptation of the proof of Proposition 8.3.)

9.7. Prove that Axiom B.2 implies Axiom B. (Hint: adapt the proof of Theorem I.16.1).

9.8. Prove that in $\widetilde{\mathcal{C}}^{op}_{C}$, every object N of form i(M) (= $\mathrm{Spec}_R(C^\infty(M))$) is reflexive: the canonical

$$N \longrightarrow \underset{\approx}{\mathrm{Hom}}_{R\text{-}Alg}(R^N, R)$$

is an isomorphism (R-Alg denotes, again, $(R \downarrow \Pi_\infty\text{-Alg}(E))$.) More generally

$$N^X \cong \underset{\approx}{\mathrm{Hom}}_{R\text{-}Alg}(R^N, R^X)$$

9.9. (Dubuc [11]). Let $\mathcal{C} = \mathcal{W} \subseteq FG\Pi_\infty$ be the category of Weil algebras over $\mathbb{R}$. Prove that the open-cover topology on it is trivial, so that $\widetilde{\mathcal{C}}^{op} = \underline{Set}^{\mathcal{C}}$. Prove that the functor i Mf $\longrightarrow \underline{Set}^{\mathcal{C}}$ given by $i(M)(W) = \hom_{\Pi_\infty\text{-}Alg}(C(M), W)$ is a "well-adapted model" in the sense that it is faithful (but not full), and satisfies Axioms A,B,C, and D of §§ 3 and 4. Prove that for this model

$$\vdash_1 \forall x \in R: (x \text{ is invertible}) \vee (x \text{ is nilpotent})$$

9.10 (Dubuc). Let $\mathcal{C}$ satisfy (9.1), and assume that $\mathbb{R} \in \mathcal{C}$. Let $N \in \widetilde{\mathcal{C}}^{op}$ be an object with no global elements $\mathbf{1} \to N$. Prove that $N = \emptyset$. (Hint: Prove that if $yC \to N$ is any generalized element, $yC = \emptyset$. Now use that the representables are topologically dense).

This should be viewed as a "Nullstellensatz".

III.10: A FIELD PROPERTY OF R, AND THE SYNTHETIC ROLE OF GERM ALGEBRAS

We consider in this § the models $\mathcal{E} = \widetilde{\mathcal{B}^{op}}$ and $\mathcal{E} = \widetilde{\mathcal{C}^{op}_C}$. We have in both cases the Axioms A,B,C, and D of §§ 3 and 4 satisfied, by Theorem 8.4 and Theorem 9.4, respectively. In particular, we know by Proposition 4.4 that $R\ (= i(\mathbb{R}) = y(C^{\infty}(\mathbb{R})))$ is a local ring object.

It turns out that in a certain sense, R is a field object, but with a field notion which is not contradictory to the fundamental Axiom 1 (cf. I §1). To state it, we remind the reader of Part II. We use $\bigwedge_{i=1}^{n} \varphi_i$ as a shorthand for the n-fold conjunction $\varphi_1 \wedge \ldots \wedge \varphi_n$, and similarly for disjunction.

<u>Theorem 10.1</u>. For $\mathcal{E} = \widetilde{\mathcal{B}^{op}}$ and $\mathcal{E} = \widetilde{\mathcal{C}^{op}_C}$, $R = \mathcal{E}$ satisfies, for each natural number n,

$$\vdash_1 \forall x_1,\ldots,x_n : \neg(\bigwedge_{i=1}^{n} (x_i = 0)) \Leftrightarrow \bigvee_{i=1}^{n} (x_i \text{ is invertible}) \tag{10.1}$$

<u>Proof</u>. We first consider the case $\mathcal{E} = \widetilde{\mathcal{B}^{op}}$. Then the category $\mathcal{E}$ has the representables as a dense class $\mathcal{A}$ of generators which is topologically dense; so we may define the satisfaction relation $\vdash$ relative to that, cf. II §§ 7 and 8. So let there be given an n-tuple of (generalized) elements of R, defined at stage $yB \in \mathcal{A}$:

$$yB \xrightarrow{x_i} R \qquad i = 1,\ldots,n,$$

and assume that

$$\vdash_{yB} \neg(\bigwedge_{i=1}^{n} (x_i = 0)).$$

Under the bijective correspondences

$$\hom_{\mathcal{E}}(yB, y(C^{\infty}(\mathbb{R}))) = \hom_{\mathcal{B}}(C^{\infty}(\mathbb{R}), B) \cong B,$$

$x_i \in_{yB} R$ corresponds to an actual element $\bar{x}_i \in B$. The ideal $(\bar{x}_1,\ldots,\bar{x}_n) \subseteq B$ is germ-determined, by Theorem 6.3; let $C = B/(\bar{x}_1,\ldots,\bar{x}_n)$, and let $\beta: B \to C$ be the canonical map. Under the change of stage $y(\beta): yC \longrightarrow yB$, the generalized elements $x_i \in_{yB} R$ all become zero, because all $\bar{x}_i$ become zero under $\beta: B \longrightarrow C$. By the assumption it follows that C is covered by the empty family, whence $C = \{0\}$. So $(\bar{x}_1,\ldots,\bar{x}_n) = B$. We consider the family of maps in $\mathcal{B}$

$$\{B \xrightarrow{\xi_i} B\{\bar{x}_i^{-1}\}^{\wedge} \mid i = 1,\ldots,n\} . \tag{10.2}$$

If $p: B \to \mathbb{R}$ is a point of B, $p(\underline{x}_1),\ldots,p(\underline{x}_n)$ generate the unit ideal of $\mathbb{R}$, hence at least one $p(\underline{x}_i)$ is invertible, so that p factors across ξ_i. Thus the family (10.2) is a covering for the open cover co-topology on $\mathcal{B}$. The change of stage $y(\xi_i)$ makes x_i invertible, since ξ_i makes $\bar{x}_i$ invertible. But this implies, by definition of $\vdash$ for disjunctive statements that

$$\vdash_{yB} \bigvee_{i=1}^{n} (x_i \text{ is invertible})$$

This proves the implication '$\Rightarrow$'. The converse implication is almost trivial:

First

$$\vdash_1 \forall x \; (x \text{ invertible} \Rightarrow \neg(x = 0)). \tag{10.3}$$

For, if $x: yB \to R$ satisfies $\vdash_{yB}$ (x is invertible), then $\bar{x} \in B$ is invertible in B. Thus, if $\beta: B \longrightarrow C$ sends $\bar{x}$ to 0, C must be the zero ring. This proves $\vdash_{yB} \neg(x = 0)$. Secondly, from (10.3), the result follows by purely logical means, since it is easy to see that (cf. (Exercise II.8.8)

$$\vdash_X (\bigvee_{i=1}^{n} \varphi_i) \Rightarrow \neg(\bigwedge_{j=1}^{n} \psi_i)$$

is implied by

$$\vdash_X \bigwedge_{i=1}^{n} (\varphi_i \Rightarrow \neg\psi_i),$$

for any X, φ_i and ψ_i.

We next consider the "Cahiers topos" $\widetilde{\mathcal{C}_C^{op}}$. The argument for the non-trivial implication '$\Rightarrow$' cannot be repeated, since we cannot in general form $B/(\bar{x}_1,\ldots,\bar{x}_n) \in \mathcal{C}_C$ when $B \in \mathcal{C}$ and $\bar{x}_i \in B$. Instead, the argument goes as follows: Let $x_i: yB \to R$ $(i = 1,\ldots,n)$ be an n-tuple of generalized elements of R, defined at stage yB, with $B = C^\infty(M) \otimes_\infty W$, say ($M$ a manifold, W a Weil-algebra). As in the proof for the $\mathcal{B}$-case, these elements correspond to n elements

$$\bar{x}_1,\ldots,\bar{x}_n \in B = C^\infty(M) \otimes_\infty W.$$

The assumption $\vdash_{yB} \neg(\bigwedge(x_i=0))$ means that no $\beta: B \to C$ in $\mathcal{C}_C$ can kill all the $\bar{x}_i$'s (except $C = \{0\}$). This in particular holds for $C = \mathbb{R}$, i.e. for the points of B. But they are of form

$$C^\infty(M) \otimes_\infty W \xrightarrow{\pi} C^\infty(M) \xrightarrow{ev_p} \mathbb{R}$$

with $p \in M$ (cf. Theorem 5.7). Since no such kills all the $\bar{x}_i$'s, it means that for each $p \in M$, at least one of the functions $\pi(\bar{x}_i) \in C^\infty(M)$ is non-zero at p. Let $U_i \subseteq M$ (for $i = 1,\ldots,n$) be the open set defined by

$$U_i := \{p \in M \mid \pi(\bar{x}_i)(p) \neq 0\} \quad ;$$

they form an open cover of M, and $\pi(\bar{x}_i)$ becomes invertible by applying the restriction $C^\infty(M) \to C^\infty(U_i)$. Also $\pi: C^\infty(U_i) \otimes_\infty W \longrightarrow C^\infty(U_i)$ reflects invertibility, and the $C^\infty(M) \otimes_\infty W \to C^\infty(U_i) \otimes_\infty W$ form a cocovering of $B = C^\infty(M) \otimes_\infty W$ in $\mathcal{C}$. The non-trivial implication ' $\Rightarrow$ ' in the Theorem now easily follows.

Corollary 10.2. With $\mathcal{E}$, R as in the Theorem, the two subobjects of R

$$[\![x \mid \neg(x=0)]\!] \quad \text{and} \quad \mathrm{Inv}(R)$$

agree.

In the following, we shall employ a notation known from topos theory (and elsewhere): if $U \rightarrowtail V$ is a subobject, $\neg(U \rightarrowtail V)$ is the subobject given as the extension.

$$[\![x \in V \mid \neg(x \in U)]\!] \quad .$$

Equivalently (in a topos), it is the unique largest subobject of V having $\emptyset$ interesection with U.

Also, if $p\colon \mathbf{1} \to V$ is any global section, it is necessarily monic, and the subobject of V it defines will be denoted $\{p\}$. Thus

$$\neg\{p\} = [\![x \in V \mid \neg(x \in \{p\})]\!] = [\![x \in V \mid \neg(x = p)]\!] .$$

We shall now use Theorem 10.1 to analyze the synthetic role of the germ algebras of form $C^\infty_p(\mathbb{R})$, where $p \in \mathbb{R}^n$ or equivalently, $p\colon \mathbf{1} \to R^n$. The algebra $C^\infty_p(\mathbb{R}^n)$ belongs to $\mathcal{B}$ but not to $\mathcal{C}_C$. Anyway, the canonical surjective $C^\infty(\mathbb{R}^n) \longrightarrow C^\infty_p(\mathbb{R})$ induces a monic map in $\mathcal{E} = \widetilde{\mathcal{B}^{op}}$

$$y(C^\infty_p(\mathbb{R}^n)) \rightarrowtail y(C^\infty(\mathbb{R}^n)) = R^n$$

as well as in $\mathcal{E} = \widetilde{\mathcal{C}^{op}}$ $(\mathcal{C} = \mathcal{C}_C$ or $\mathcal{C}_s)$,

$$y_o(C^\infty_p(\mathbb{R}^n)) \rightarrowtail y(C^\infty(\mathbb{R}^n)) = R^n \qquad ;$$

the subobject thus defined, we shall temporarily denote $(R^n)_p$. But it need no special notation because of the remarkable

Theorem 10.3. For $\mathcal{E} = \widetilde{\mathcal{B}^{op}}$ as well as for $\mathcal{E} = \widetilde{\mathcal{C}^{op}_C}$ and $\mathcal{E} = \widetilde{\mathcal{C}^{op}_S}$, the two subobjects of R^n

$$\neg\neg\{p\} \quad \text{and} \quad (R^n)_p \tag{10.4}$$

agree, for any $p: \mathbf{1} \to R^n$.

We shall first prove

Proposition 10.4. Let $\mathcal{C}$ be a full subcategory of $\mathcal{B}$, with $\mathbb{R} \in \mathcal{C}$, and with some subcanonical Grothendieck (co)-topology for which the forgetful functor $R: \mathcal{C} \longrightarrow \underline{Set}$ is a sheaf, and where $\{0\}$ (the one-element $\mathbb{T}_\infty$-algebra) is covered by the empty family. Then for any $B \in \mathcal{C}$ and any $b \in_{yB} R$, the following conditions are equivalent in $\widetilde{\mathcal{C}^{op}}$:

(i) $\vdash_{yB} \neg(b$ is invertible)

(ii) for every point $p: B \to \mathbb{R}$, $p(b) = 0$

(iii) the $\mathbb{T}_\infty$-homomorphism $b: C^\infty(\mathbb{R}) \longrightarrow B$ factors across $C^\infty_o(\mathbb{R})$.

(We identify notationally the equivalent data $b: C^\infty(\mathbb{R}) \to B$, $b \in B$, $b \in_{yB} R$.)

Proof. Assume (i). For every $C \in \mathcal{C}$ which is not covered by the empty family, any $\varphi: B \to C$ takes b to a non-invertible element in C; now $\mathbb{R} \in \mathcal{C}$, and is not covered by the empty family. So for any $p: B \to \mathbb{R}$ $p(b)$ is non-invertible in $\mathbb{R}$. But in $\mathbb{R}$, "non-invertible" means "equal zero". This proves (ii). Assume (ii). To prove (iii), let $f \in C^\infty(\mathbb{R})$ belong to the kernel of

$C^\infty(\mathbb{R}) \to C_0^\infty(\mathbb{R})$, so f vanishes on some open set U around 0. We should prove that $b: C^\infty(\mathbb{R}) \to B$ takes f to 0. It suffices, since B is germ-determined, to see that for any point p of B, $b(f)_p = 0$. Now U and $\mathrm{Inv}(\mathbb{R})$ is an open covering of $\mathbb{R}$. So the point $p \circ b$ of $C^\infty(\mathbb{R})$ must factor through either $C^\infty(\mathbb{R}) \longrightarrow C^\infty(U)$ or $C^\infty(\mathbb{R}) \to C^\infty(\mathrm{Inv})$. In the first case, $b(f)_p$ is zero, since f goes to zero in $C^\infty(U)$. The second case does not occur, since it would imply that

$$C^\infty(\mathbb{R}) \xrightarrow{\;b\;} B \xrightarrow{\;p\;} \mathbb{R} \tag{10.5}$$

sends $\mathrm{id}_{\mathbb{R}}$ into an invertible element in $\mathbb{R}$, so $p(b) \in \mathbb{R}$ is invertible, contrary to (ii). This proves (iii). Finally, assume (iii), and assume that $\varphi: B \to C$ has $\varphi(b)$ invertible in C. Then $\varphi \circ b$ factors across $C^\infty(\mathbb{R}) \longrightarrow C^\infty(\mathrm{Inv}(\mathbb{R}))$. By assumption, it also factors across $C^\infty(\mathbb{R}) \longrightarrow C_0^\infty(\mathbb{R})$, hence over their push-out in $\mathcal{B}$. But this pushout clearly has no points, hence is $\{0\}$, so that $C = \{0\}$, hence is covered by the empty family. This proves (i).

The Proposition has the following Corollary.

Corollary 10.5. If $\mathcal{C}$ is as in the Proposition, then in $\mathcal{E} = \widetilde{\mathcal{C}^{op}}$, the subobject $[\![x \in R \mid \neg(x \text{ is invertible})]\!]$ is represented (possibly from the outside) by the germ algebra $C_0^\infty(\mathbb{R})$. This in particular applies to $\mathcal{E} = \widetilde{\mathcal{B}^{op}}$, $\mathcal{E} = \widetilde{\mathcal{C}_C^{op}}$, and $\mathcal{E} = \widetilde{\mathcal{C}_S^{op}}$.

Proof of the theorem. Without loss of generality, we may assume $p = \underline{0} \in \mathbb{R}^n$. Then

$$\begin{aligned} \neg\neg\{0\} &= [\![x \in R^n \mid \neg\neg(\bigwedge_{i=1}^{n} (x_i = 0))]\!] \\ &= [\![x \in R^n \mid \neg(\bigvee_{i=1}^{n} x_i \text{ is invertible})]\!], \text{ by Theorem 10.1} \\ &= [\![x \in R^n \mid \bigwedge_{i=1}^{n} \neg(x_i \text{ is invertible})]\!], \end{aligned} \tag{10.6}$$

the last equality for logical reasons. If we denote by R_o the subobject of R represented by $C_o^\infty(\mathbb{R})$, then we get, by Corollary 10.5 that (10.6), in turn, equals the subobject

$$\bigcap_{i=1}^{n} R \times \ldots \times R_o \times \ldots \times R \tag{10.7}$$

(R_o as the i'th factor). Now the functor $y_o : \mathcal{B}^{op} \longrightarrow \mathcal{E}$ is left exact, and so is the reflection functor $\wedge : FG\mathbb{T}_\infty^{op} \longrightarrow \mathcal{B}^{op}$. The object (10.7) is therefore represented by $\hat{C}$ where $C \in FG\mathbb{T}_\infty$ is the $\mathbb{T}_\infty$-algebra

$$C = C^\infty(\mathbb{R}^n)/(I_1 + \ldots + I_n),$$

where I_i is the ideal generated by the functions $f(x_1,\ldots,x_n)$ which only depend on the i'th variable, and as a function of x_i vanish in a neighbourhood of 0.

On the other hand $(R^n)_{\underline{0}}$ is by definition represented by the $\mathbb{T}_\infty$-algebra $C^\infty(\mathbb{R}^n)/J(\underline{0})$, where $J(\underline{0})$ is the ideal of functions $\mathbb{R}^n \to \mathbb{R}$ vanishing in some neighbourhood of $\underline{0} \in \mathbb{R}^n$. This is a germ-determined ideal. The result will follow if we can prove $J(\underline{0})$ equal to the germ radical of $I_1 + \ldots + I_n$. The non-trivial part of this is proving

$$(I_1 + \ldots + I_n)^\wedge \supseteq J(\underline{0}), \tag{10.8}$$

the converse inclusion being clear. So let $f \in J(\underline{0})$. To prove that $f \in (I_1 + \ldots + I_n)^\wedge$ means proving, for all $p \in \mathbb{R}^n$, that

$$f_p \in (I_1 + \ldots + I_n)_p = (I_1)_p + \ldots + (I_n)_p . \tag{10.9}$$

For $p = \underline{0}$, this is clear, since f has zero germ at $\underline{0}$. For $p \neq 0$, $p = (x_1,\ldots,x_n)$ with (say) $x_j \neq 0$. So $(I_j)_p$ is the unit ideal, so (10.9) again holds. This proves (10.8), hence the Theorem.

We may introduce an "apartness" relation $\#$ on R^n by putting, for $\underline{x}$ and $\underline{y}$ generalized elements of R^n defined at same stage,

$$\underline{x} \# \underline{y} \text{ iff } \bigvee_{i=1}^{n} (x_i - y_i \text{ is invertible}). \tag{10.10}$$

In the presence of the field property of Theorem 10.1, we of course have $\underline{x}\#\underline{y}$ iff $\neg(\underline{x} = \underline{y})$. Combining Theorem 10.1 and Theorem 10.3, we therefore have, for $\mathcal{E} = \widetilde{\mathcal{B}^{op}}$, $\mathcal{E} = \widetilde{\mathcal{C}_C^{op}}$, and $\mathcal{E} = \widetilde{\mathcal{C}_S^{op}}$

<u>Theorem 10.6</u>. For any $p\colon \mathbf{1} \longrightarrow R^n$, the three subobjects of R^n

$$[\![\underline{x} \in R^n \mid \neg\neg(\underline{x} = p)]\!], \quad [\![\underline{x} \in R^n \mid \neg(\underline{x} \# p)]\!], \text{ and } \quad y_o(C_p^\infty(\mathbb{R}^n))$$

agree.

<u>EXERCISES</u>

10.1 [30], cf. also [66]. Prove that a ring object which satisfies the conclusion of Theorem 10.1 has the property (for each n):

$$\vdash_1 \text{Any injective linear } R^n \to R^n \text{ is bijective.}$$

(Hint: formulate a notion of "k-tuple of independent vectors in R^n". Use the conclusion of Theorem 10.1 to bring any $n \times n$ matrix with n independent columns onto triangular form (by the usual sweeping method) with invertible elements in the diagonal.)

10.2. Let $\mathcal{E}$ be a category with a Grothendieck pretopology. Prove that for any $f\colon A \to B$ in $\mathcal{E}$, we have

$$\vdash_1 \forall a_1, a_2 \in A \times A\colon \; \neg(a_1 = a_2) \Rightarrow \neg(f(a_1) = f(a_2)). \tag{10.11}$$

Penon [64] has suggested consideration of the "germ-neighbourhood of the diagonal" of any object M in a topos, as being

$$[\![(x,y) \in M \times M \mid \neg\neg(x=y)]\!] \subseteq M \times M$$

Prove that if $M = R^n$ in $\widetilde{\mathcal{B}^{op}}$ or $\widetilde{\mathcal{C}_C^{op}}$, then it contains $[\![(x,y) \in M \times M \mid x \sim_k y]\!]$, for any k. Define the notion of "germ-monad" around an element in analogy with the notion of 'k-monad'. Then (10.11) plays the role of Proposition I.17.5.

10.3. (Grayson; Penon). Call a subobject U of an object E in a topos <u>open</u> if

$$\vdash_1 \forall x \in U \quad \forall y \in E: y \in U \vee \neg(y = x).$$

Prove that, with $\mathcal{E}$,R as in Theorem 10.1, for any $\underline{z} \in R^n$, the object $\{\underline{x} \in R^n \mid \neg(\underline{x} = \underline{z})\}$ is open in this sense.

10.4 (Reyes (?)). Prove that for the models $\mathcal{E}$,R considered in Theorem 10.1, we have, (for each n = 1,2,...)

$$\vdash_1 \forall x_1,\ldots,x_n: (\sum x_i^2 = 0) \Rightarrow (\neg\neg((x_1,\ldots,x_n) = (0,\ldots,0)))$$

(Hint: Use Theorem 10.1 and Exercise 4.1.).

We consider first $\mathcal{E} = \widetilde{\mathcal{C}^{op}}$, where $\mathcal{C} \subseteq \mathcal{B}$ is any category satisfying the Assumption of §9. We let $R = y(C^\infty(\mathbb{R}))$ and $H = y_o(C^\infty(\mathbb{H}))$; H is a subobject of R. In the notation of §9, $H = i'(\mathbb{H})$.

<u>Theorem 11.1</u>. The subobject $H \subseteq R$ is formal-étale.

<u>Proof</u>. Let W be a Weil algebra. Write $\bar{W}$ for $\mathrm{Spec}_R(W))$. We must prove

$$\begin{array}{ccc} H^{\bar{W}} & \overset{i^{\bar{W}}}{\rightarrowtail} & R^{\bar{W}} \\ {\scriptstyle \pi}\downarrow & & \downarrow{\scriptstyle \pi} \\ H & \underset{i}{\rightarrowtail} & R \end{array}$$

to be a pull-back. It suffices to test for any $C \in \mathcal{C}$ and any $\check{\varphi}: yC \to R^{\bar{W}}$ that if $\pi \circ \check{\varphi}$ factors through H, then $\check{\varphi}$ itself factors through $H^{\bar{W}}$. Taking exponential adjoints, we are given a $\varphi: yC \times \bar{W} \longrightarrow R$, whose restriction to $yC \rightarrowtail yC \times W$ factors through H. We must prove that φ itself factors through H. Since all objects under consideration are representable, we are looking at the situation

$$\begin{array}{ccc} C^\infty(\mathbb{R}) & \twoheadrightarrow & C^\infty(\mathbb{H}) \\ {\scriptstyle \varphi}\downarrow & \swarrow & \downarrow \\ C \otimes_\infty W & \longrightarrow & C \end{array} \qquad (11.1)$$

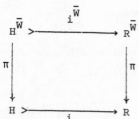

where we must provide the diagonal fill-in, as indicated. The kernel of the top map is the ideal I of functions vanishing on $\mathbb{H}$. We must prove that $f \in I$ implies $\varphi(f) = 0$. By the Assumption of §9, $C \otimes_\infty W \in \mathcal{C}$, and is thus W-determined. So it suffices to prove that for any $\mathbb{T}_\infty$-homomorphism $\lambda: C \otimes_\infty W \longrightarrow V$, where V is a Weil-algebra, $\lambda(\varphi(f)) = 0$. Let $\pi: V \to \mathbb{R}$ be the unique point of V. Then $\pi \circ \lambda \circ \varphi$ is a point of $C^\infty(\mathbb{R})$, so is (by fullness) given as "evaluation at $p \in \mathbb{R}$" for some unique $p \in \mathbb{R}$ ("$\lambda \circ \varphi$ is centered at p"). There is a bijective correspondence between points of $C \otimes_\infty W$ and points of C, and from this easily follows that p cannot be < 0, hence $p \in \mathbb{H} \subseteq \mathbb{R}$. Let $\rho: V \to \mathbb{R}$ be any $\mathbb{R}$-linear map. By Proposition 5.12,

$$\rho(\lambda(\varphi(f))) = P(\frac{\partial}{\partial x})(f)(p)$$

for some polynomial P. But since $p \in \mathbb{H}$ and f vanishes on $\mathbb{H}$, all the derivatives of f vanish at p, so $P(\frac{\partial}{\partial x})(f)(p) = 0$. Since this holds for any ρ, $\lambda(\varphi(f)) = 0$. The theorem then follows.

Let $\mathbb{N} = \{x \in \mathbb{R} \mid x < 0\} \subseteq \mathbb{R}$, and let $N = y(C^\infty(\mathbb{N}))$. Then $N \subseteq R$ is a subobject. We have

<u>Theorem 11.2</u>. $(\neg N) = H$.

<u>Proof</u>. To prove $H \subseteq \neg N$ means proving $H \cap N = \emptyset$. Clearly, $H \cap N$ can have no <u>global</u> points $\mathbf{1} \longrightarrow H \cap N$, (since such, by fullness of $Mf' \longrightarrow \mathcal{E}$, would come from an element in $\mathbb{R}$ being at the same time in $\mathbb{H}$ and in $\mathbb{N}$). If now $yC \longrightarrow H \cap N$ is a generalized element, yC can have no global points, hence $C \in \mathcal{C}$ is cocovered by the empty family. We conclude that $H \cap N$ is covered by the empty family, hence is $\emptyset$. (Note that the last part of the argument was really the argument required for "Dubuc's Nullstellensatz", Exercise 9.10).

Conversely, assume $x \in_{yC} R$ has $\vdash_{yC} \neg(x \in N)$. We must prove that the map $\bar{x}: C^{\infty}(\mathbb{R}) \to C$ factors across $C^{\infty}(\mathbb{H})$, i.e. annihilates the ideal of functions f vanishing on $\mathbb{H}$. This is almost as in the proof of Theorem 12.1; it suffices, by W-determinedness of C, to see that for any $\lambda: C \to V$ with V a Weil-algebra, $f \in I$ implies $\lambda(\bar{x}(f)) = 0$. The point $\in \mathbb{R}$ at which $\lambda \circ \bar{x}$ is centered cannot be < 0, since this would imply that $\lambda \circ \bar{x}$ factors across $C^{\infty}(\mathbb{N})$ (by formal étaleness of $N \subseteq R$), i.e.that

$$\vdash_{yV} x \circ y(\lambda) \in N$$

contrary to $\vdash_{yC} \neg(x \in N)$. So $\lambda \circ \bar{x}$ is centered at some $p \in \mathbb{H}$. The proof now proceeds exactly as the proof of Theorem 11.1.

We may now define binary relations $\leq$ and $<$ on R: we say, for generalized elements x and y of R (defined at same stage, yC, say) that

$$\vdash_{yC} x \leq y \quad \text{if} \quad \vdash_{yC} (y-x) \in H$$

and

$$\vdash_{yC} x < y \quad \text{if} \quad \vdash_{yC} (x-y) \in N.$$

Alternatively, we may define $\textcircled{\leq} \rightarrowtail R \times R$ as the extension

$$[\![(x,y) \in R \times R \mid x \leq y]\!]$$

It can be constructed as the pull-back

$$\begin{array}{ccc} \textcircled{\leq} & \longrightarrow & R \times R \\ \downarrow & & \downarrow \\ H & \rightarrowtail & R \end{array} \qquad \psi = \text{"twisted minus"}$$

where ψ has description $(x,y) \longmapsto y-x$. Similarly for $\textcircled{<}$.

Clearly $\textcircled{<}$ is contained in $\textcircled{\leq}$, i.e.

$$\vdash_1 \forall x\colon x < 0 \;\Rightarrow\; x \leq 0.$$

We have the following properties:

<u>Theorem 11.3</u>. The relations $<$ and $\leq$ are transitive, and compatible with the ring structure on R. Furthermore,

i) $\vdash_1$ $\forall d\colon$ d nilpotent $\quad 0 \leq d$

ii) $\vdash_1$ $\forall x\colon x < 0 \Rightarrow x$ invertible

iii) $\vdash_1$ $\forall x\colon x$ invertible $\Rightarrow (x < 0) \vee (-x < 0)$

iv) $\vdash_1$ $\forall x\colon \neg(x < 0) \Leftrightarrow 0 \leq x$

v) $\vdash_1$ $\forall x,y\colon (0 < x) \wedge (x \leq y) \Rightarrow 0 < y.$

<u>Proof</u>. Those assertions that deal only with $<$ are straightforward, due to the fact that $\mathbb{N} \subseteq \mathbb{R}$ is a submanifold, and $i\colon \mathbf{Mf} \to \mathcal{E}$ is full and preserves transversal pull-backs. Also, i) follows from Theorem 11.1 and iv) from Theorem 11.2; v) follows from the fact that $i'\colon \mathbf{Mf} \longrightarrow \mathcal{E}$ is full and preserves products $M \times B$ where M is boundaryless, cf. Theorem 9.5. So the only delicate point is proving $\leq$ compatible with the ring structure, and transitive (note that $\mathbb{H} \times \mathbb{H}$ does not belong to $\mathbf{Mf}'$). For this, we need the following "Positiv-stellen-satz":

<u>Lemma 11.4</u>. Let $C \in \mathcal{C}$ have a presentation $q : C^\infty(\mathbb{R}^m) \twoheadrightarrow C$. Let $g \in C^\infty(\mathbb{R}^m)$, and let $\bar{g} = q(g) \in C$. If J is the kernel of q, the following conditions are equivalent:

(i) $\quad g$ maps $Z(J)$ into $\mathbb{H}$.

(ii) $\vdash_{yC} 0 \leq \hat{g}$ (where $\hat{g}: C^{\infty}(\mathbb{R}) \to C$ is the homomorphism sending $id_{\mathbb{R}}$ to $\bar{g}$, and where we omit $y: \mathcal{C}^{op} \longrightarrow \mathcal{E}$ from notation).

Proof. Assume (i). To prove (ii) means to prove that $\hat{g}$: $C^{\infty}(\mathbb{R}) \longrightarrow C$ factors across $C^{\infty}(\mathbb{R}) \longrightarrow C^{\infty}(\mathbb{H})$, i.e. that $\hat{g}$ annihilates the ideal I of functions vanishing on $\mathbb{H}$. Since C is W-determined, it suffices to see that $\hat{g}$ annihilates the ideal I' of functions vanishing in some open subset of $\mathbb{R}$ containing $\mathbb{H}$ (for, the W-radical of I' is I, by Exercise 9.3). Let $f \in I'$, say f vanishes on $U \supseteq \mathbb{H}$. Then

$$Z(J) \subseteq g^{-1}(\mathbb{H}) \subseteq g^{-1}(U) \subseteq \mathbb{R}^{m},$$

and $f \circ g$ vanishes on $g^{-1}(U)$. Since $g^{-1}(U)$ is open, the germ of $f \circ g$ at any $p \in Z(J)$ is zero. This implies $\hat{g}(f)_p \in C_p$ is zero, for any point p of C, and since C is germ-determined, $\hat{g}(f) = 0$.

Conversely, assume (ii), and let $p \in Z(J)$. Then $p: C^{\infty}(\mathbb{R}^{m}) \longrightarrow \mathbb{R}$ factors across C, as $\bar{p}$, say. If $g(p) \notin \mathbb{H}$, $g(p) < 0$, and

$$C^{\infty}(\mathbb{R}) \xrightarrow{\hat{g}} C \xrightarrow{\bar{p}} \mathbb{R}$$

may be interpreted as a generalized element $\in_{y\mathbb{R}} R$ obtained by base change from $\hat{g}$ along $\bar{p}$. But $\vdash_{yC} 0 \leq \hat{g}$, hence $\vdash_{y\mathbb{R}} 0 \leq \hat{g}$. On the other hand, $g(p) < 0$ implies $\vdash_{y\mathbb{R}} \hat{g} < 0$. These two things are incompatible, by iv) of the theorem. So $g(p) \notin \mathbb{H}$ is incompatible with the assumption. This proves the Lemma.

To finish the proof of the remaining parts of the theorem: transitivity of $\leq$, and compatibility with the ring structure, we utilize the Lemma. To prove $0 \leq x$ and $0 \leq y$ implies $0 \leq x+y$, consider

$$yC \underset{\bar{b}}{\overset{\bar{a}}{\rightrightarrows}} H \subseteq R$$

To prove $\vdash_{yC} 0 \leq (\bar{a}+\bar{b})$, take a presentation of C as in the proof of the Lemma; $\bar{a}$ and $\bar{b}$ are represented by $a,b \in C^{\infty}(\mathbb{R}^m)$. By the Lemma, it suffices to see that $(a+b)(p) \geq 0$ for any $p \in Z(J)$. But $a(p) \geq 0$, $b(p) \geq 0$ by the Lemma, whence $a(p)+b(p) \geq 0$, whence the result. The other remaining parts of the theorem are proved similarly.

Let us remark the identity of the following two subobjects of R:

$$i'([0,1]) = [0,1] \quad (= [\![x \in R \mid 0 \leq x \leq 1]\!]). \tag{11.2}$$

The former is $y_o(C^{\infty}[0,1])$, the latter is the intersection of H and $H' = (-H) + 1$; H and H' are represented by the two $\mathbb{T}_{\infty}$-algebras $C^{\infty}(\mathbb{H})$ and $C^{\infty}(\mathbb{H}')$ (where $\mathbb{H}' = \{x \mid x \leq 1\}$). Since $y_o : \mathcal{B}^{op} \longrightarrow \mathcal{E}$ preserves all limits that exist, we see that $H \cap H'$ is represented by

$$C^{\infty}(\mathbb{R})/(I+I')$$

where I (respectively I') is the ideal of functions vanishing on $\mathbb{H}$ (respectively on $\mathbb{H}'$). This ideal equals the ideal J of functions vanishing on $[0,1]$; for, clearly $I+I' \subseteq J$, and if $f \in J$, we may find a partition of unity φ, ψ subordinate to the covering $(-\infty, 2/3)$, $(1/3, \infty)$. Then

$$f = \varphi \cdot f + \psi \cdot f,$$

and clearly $\varphi \cdot f \in I$, $\psi \cdot f \in I'$.

We shall finally consider the integration theory of I §13 for the specific model $\mathcal{E} = \widetilde{\mathcal{C}_C^{op}}$ (the "Cahiers topos", $\mathcal{C}_C$ being

$\mathbb{T}_\infty$-algebras of form $C^\infty(M) \otimes_\infty W$). The fact from analysis which enters here is the following: given a smooth

$$M \times [0,1] \xrightarrow{f} W = \mathbb{R}^m ; \tag{11.3}$$

we form out of this another smooth map, denoted $\int f$

$$M \times [0,1] \xrightarrow{\int f} W , \tag{11.4}$$

by putting

$$(\int f)(m,t) := \int_0^t f(m,s)ds$$

(the integration of functions with values in $W = \mathbb{R}^m$ takes place in each of the m $\mathbb{R}$-factors separately, as usual).

With $E = \widetilde{C}^{op}_C$, $R = i(\mathbb{R}) = y(C^\infty(\mathbb{R}))$, and with the relation $\leq$ on R derived from $C^\infty(\mathbb{H})$, as above, we have

<u>Theorem 11.5</u>. The Integration Axiom of I §13 holds.

<u>Proof</u>. Let $f: yC \to R^{[0,1]}$ be a generalized element. By assumption, $C = C^\infty(M) \otimes_\infty W$ for some $M \in Mf$ and W a Weil algebra. So f comes in the following equivalent conceptual disguises (where m is the dimension of W as a vector space):

$$\frac{y(C^\infty(M)) \times y(W) \cong y(C^\infty(M) \otimes_\infty W) \longrightarrow R^{[0,1]}}{\frac{y(C^\infty(M)) \times [0,1] \longrightarrow R^{y(W)}}{\frac{i(M) \times i'([0,1]) \longrightarrow R^{y(W)} \cong R \otimes W = R^m}{\frac{i'(M \times [0,1]) \longrightarrow R^m}{M \times [0,1] \longrightarrow \mathbb{R}^m \text{ (smooth)},}}}} \tag{11.5}$$

the first passage by twisted exponential adjointness, the second by Axiom 1^W (recall by (9.3), $yW = \text{Spec}_R W$), as well as by (11.2), the third because i' preserves certain products (cf. Theorem 9.5), and the fourth because i' is full (cf. Theorem 9.6). This is the data in (11.3). Pass to (11.4) by the prescription given there, and with (11.4), go backwards through (11.5). Thus, we obtain a new element of $R^{[0,1]}$, defined at stage yC, again; this element satisfies the requirement for y in the integration axiom, because of two facts:

1) the function $\int f$ defined in (11.4) satisfies

$$(\int f)'(m,t) = f(m,t) \quad \text{and} \quad (\int f)(m,0) = 0$$

(where the prime denotes derivative w.r.to t, keeping m fixed),

2) i' preserves differentiation. This was proved for functions $\mathbb{R} \to \mathbb{R}$ in Theorem 3.2 (or better (3.3)), but the proof for functions $[0,1] \to \mathbb{R}$ is the same. We leave the details (i.e. keeping tack of the identifications) to the reader, (or cf.[44]).

EXERCISES

11.1 (Reyes). Use that the Whitney Theorem on smooth even functions (quoted in §9) is smooth in parameters to prove that in $\widetilde{C}_C^{op}$, the diagram

$$R^H \xrightarrow{R^{sq}} R^R \underset{R^{id}}{\overset{R^{-id}}{\rightrightarrows}} R^R$$

is an equalizer, or alternatively, that 'internal Whitney Theorem' holds:

$$\vdash_1 \forall f \in R^R\ (\forall x \in R\colon f(x) = f(-x)) \Rightarrow \exists!\ g\colon H \to R\colon$$
$$(\forall x \in R\colon f(x) = g(x^2)).$$

11.2. Assume $i: \mathbf{Mf} \longrightarrow \mathcal{E}$ satisfies Axiom A and C of §§ 3 and 4, and define $\textcircled{<} \rightarrowtail R \times R$ as in Exercise 4.1. Prove that this is the extension $[\![(x,y) \in R \times R \mid x < y]\!]$ with $<$ defined as in the present §.

Also prove, on basis of Axiom A and C alone, that

$$\vdash_1 \forall x \in R: 0 < x \Leftrightarrow \exists y \in \mathrm{Inv}(R): y^2 = x.$$

11.3. Prove that if $\mathcal{C} \subseteq \mathcal{B}$ is a subcategory satisfying the Assumption of §9, then the relation $\leq$ on R as defined in the present § can be described entirely in terms of the ring structure on R, as follows:

$$\vdash_1 \forall x \in R: x \leq 0 \Leftrightarrow \neg(\exists y \in \mathrm{Inv}(R): y^2 = x).$$

(Hint: use Exercise 11.2 and Theorem 11.2).

11.4. Prove that for the model considered in Exercise 9.9 we have non-constant maps $f: R \longrightarrow R$ with $f' \equiv 0$.

LOOSE ENDS

There are several developments, related to the content of the present book, which were not treated in it, due to lack of time or space, or lack of comprehension on my side. I would have liked to include something about

1. Formulation and application of a synthetic inverse function theorem. A formulation has been given by Penon [64]: he utilizes 1) that an inverse function theorem expresses that a germ is invertible if its 1-jet is invertible 2) a synthetic formulation of the germ notion is implied by his Theorem (Theorem III.10.3) (and, of course, a synthetic formulation of the jet-notion is implied in the foundations of synthetic differential geometry).

In particular, a synthetic version of the Preimage theorem (as quoted in III §3) is still lacking.

2. Differential equations. A correct formulation of the classical existence - and uniqueness theorems for flows for vector fields is still lacking. The problem is that solutions (flows) only should exist locally, (cf. e.g. Exercise I.8.7). Maybe the Penon germ notion implied in Theorem III.10.3 will provide a formulation, which is true in some og the models, and strong enough to carry some synthetic theory.

3. Calculus of variations. The problems and perspectives are related to those mentioned for differential equations. However, a gros topos designed for this purpose, and for the purpose of considering differential forms as quantities, exists implicitly

in work of K.T. Chen [4], as Lawvere has pointed out.

4. Methods of differential algebra in synthetic context. A start on this has been made by M. Bunge [3], who considers the ring R^R as a differential ring, and relates it to a notion "ring of differential line type" (proving in particular that the generic differential ring is such).

5. Algebraic topology from the differentiable viewpoint. Since we are considering differential forms, we should also ask to what extent, or under which further axioms, deRham cohomology comes out right. This and related questions are being studied in a forthcoming M.Sc. thesis by L.Bélair (U. de Montréal).

Finally, a certain "red herring" must be mentioned. We consider the picture in Exercise I.1.2 but in one dimension higher; thus we are asking for the intersection of a unit sphere and one of its tangent planes. Taking the center of the sphere in (0,0,1), the xy plane is a tangent plane, and the intersection is

$$[\![(x,y) \in R^2 \mid x^2 + y^2 = 0]\!] \subseteq R \times R . \qquad (*)$$

Is this one of the infinitesimal objects we already know? Clearly not; in certain models, it even contains elements far away from (0,0), like in $\underline{\mathrm{Set}}^{\mathrm{FP}\mathbb{T}_{\mathbb{R}}}$, where it contains for instance the element, defined at stage $y(\mathbb{C})$,

$$y(\mathbb{C}) \xrightarrow{(1,i)} R \times R,$$

since $1^2 + i^2 = 0$.

So some reality properties of R must be postulated to get (*) to be infinitesimal. Does there exist an ε-stable geometric theory $\mathcal{T}$ of $\mathbb{R}$-algebras (say), such that the algebra $A = \mathbb{R}[X,Y]/(X^2+Y^2)$ has a universal $\mathbb{R}$-algebra homomorphism $A \to B$ into a $\mathcal{T}$-model B, in such a way that B is a Weil algebra (note: A is not a Weil algebra, being infinite dimensional as a vector space).

If there is such $\mathcal{T}$ and B, B cannot be too small. In particular, $\mathrm{Spec}_R(B)$ must be strictly larger than $D \times D$; this follows from Exercise I.4.7.

On the other hand, the models $\tilde{\mathcal{B}}^{op}$ and $\tilde{C}_c^{op}$ of III §10 have the property that (*) is contained in the "¬¬-monad" $[\![\underline{x} \in R^2 \mid \neg\neg(\underline{x}=\underline{0})]\!]$ of $\underline{0} \in R^2$. (Reyes; cf. Exercise III.10.4).

HISTORICAL REMARKS

I.1 and I.2. The formula $R^D \simeq R \times R$ appeared in a lecture by Lawvere in 1967 ("Categorical Dynamics"), and in the form here, with the explicitely given isomorphism α, in [31], 1976. With this, its power was recognized, and the present development of the theory took form. The paper [31] was not the first one inspired by Lawvere's 1967 talk. In 1972, Wraith, in an unpublished manuscript, considered the construction of the Lie algebra of a group object in "categorical dynamics" terms, and for this purpose, considered what we have called "Property W"; see the historical remarks for I.9. A summary, written in 1979, of Lawvere's 1967 talks, exists, [50].

In [31] and later, one called rings satisfying Axiom 1 (or Axiom 1', 1",..) ring objects of line type. The terminology "rings satisfying the Kock-Lawvere axiom" is now also used.

The naive verbal formulations (supplementing the diagrammatic ones) became current in the late 70's.

I.3 - I.5. This is largely from [32]. The consideration of the D_k's was suggested by G. Wraith in 1976.

I.6. The notion of infinitesimal linearity in some sense goes back to [10]. In the present form and context, it emerged in discussion in 1977 between Reyes, Wraith, and myself, where we also considered many of the infinitesimal objects and their relationship. The consideration of a k-neighbour relation occurs in classical scheme-theory. Joyal advocated its use in synthetic context, cf. historical remarks to I.18.

I.7 and I.8. The consideration of M^D as tangent bundle, and the notion of vector field in terms of this was explicite in Lawvere's 1967 talk, and in Wraith's investigations of 1972-74. In particular, Lawvere emphasized in 1967 how the necessary conceptual transformations (between vector fields, infinitesimal transformations, and tangent vectors at $id \in M^M$) make also the notion of cartesian closed category necessary.

I.9. The content here is from [71]; the idea of Property W to define Lie bracket is due to the unpublished work of Wraith in the early 70's.

I.10. This is mainly from [34], 1978.

I.12. This is mainly from [38], 1980.

I.13. This is mainly from [44], 1979.

I.14 and I.15. The form notion of Definition 14.2 is from [45]. We considered Stokes' theorem from the viewpoint, as presented here, that it is tautologically true for infinitesimal chains.

I.16. Weil algebras were considered by Bourbaki and his student, A. Weil, in a remarkable paper by the latter, [79], "Théorie des points proches sur les varietes differentiables", for purposes related to those of the present book. However, they did not develop the viewpoint. A closely related start was made by Emsalem [16], but again not developed. The use of Weil algebras is the present approach is due to Dubuc [11]. Axiom 1^W was formulated in [36].

I.17. The formal-étaleness notion introduced here is akin to that of algebraic geometry. Essentially it was introduced into synthetic context in [42]. We used this étaleness notion to introduce one notion of 'manifold' synthetically. Another one was introduced in [35] , and the one of the present § is akin to that. A completely new idea for a good manifold notion is due to Penon, cf. historical remarks for III.10.

I.18. The simplicial object (18.1) was advocated by Joyal, who, together with Bkouche, used it for defining differential forms (unpublished) and de Rham cohomology. The object $M_{(1)}$ is presumably in suitable contexts (local ringed spaces) and for suitable M equal to Grothendieck's "1st neighbourhood of the diagonal", [23], which also in the context of smooth manifolds has been investigated to a considerable extent, [58], [46], and others.

The correspondence between the group valued froms here and 'classical' Lie algebra valued forms is from [37] .

I.19. Atoms were first used, and examples of them found, in [42]. In particular, we proved that even in certain subtoposes of the ring classifier (Zariski topos, étale topos), the object D is an atom. We used the atom property in connection with formal-étale descent. The strong étaleness notion is stronger, and inspired by recent work of Joyal, and of Coste-Michon [7].

I.20. It was Lawvere who in 1979 pointed out that, given an atom J, the right adjoint of $(-)^J$ should be studied and utilized, in particular so as to be able to view differential forms as quantities. The content of §20 is mainly due to him. Much research is still to be done. What do $\Lambda^n(V)$ or V_{D^n} look like in concrete models? What are the categorical properties of atoms J, and the functors $(-)_J$? The latter question is genuinely category-theoretic, since categorical logic, as we know it now, is unable to talk about $(-)_J$.

It would also be interesting to see whether the kind of differential forms considered in I.18 can be reinterpreted as quantities: Joyal has conjectured that the functor $M \mapsto M_{(1,\ldots,1)}$

likewise should have a right adjoint. But at least as presented here, the functor is only defined on formal manifolds M; a good general definition of $M_{(1)}$ is still lacking.

II.1 - II.9. In its first stages of its present development, synthetic differential geometry was formulated mainly in diagrammatic terms. The elementwise mode of expressing its notions and reasonings developed later, alongside with an explicitation in logical terms of the semantics of generalized elements.

Generalized elements, or elements defined at different stages, have a long tradition in geometry, cf. Introduction, and the remarks at the end of III §1. The precise category theoretic formulation of the notion is probably due to the group around Grothendieck, who also emphasized the role of the Yoneda Lemma in this connection. They, however, refused to give an explicite semantics for logical formulas talking about such elements (except when the formulas were purely equational). Such semantics was advocated by Joyal in 1972, and possibly by others too. The properties and applicability of this semantics was tested through the works [40], [30] , [67] , and other places. We used the term 'Kripke-Joyal semantics'. Now we call it sheaf semantics (implicitely, it was present in sheaf theory all the time).

The good logical properties of toposes were stressed much earlier (implicitely by Lawvere in [48] in 1964) and in 1969-70 with the development of the theory of elementary toposes by Lawvere and Tierney [21] . An explicite semantics for suitable formal languages, exploiting theses logical properties, was given by W. Mitchell [59] and J. Bénabou [2]; they gave the notion of extension of a formula the leading role. For toposes, the semantics in terms of elements is equivalent to the semantics in terms of extensions (cf. e.g. [63]).

The group around Bénabou (notably M. Coste) proved a formal metatheorem, according to which a certain natural form of intuitionistic logic is sound (=only yields valid results) for this interpretation. This metatheorem may be seen as the justification for the fact that we do not 'revisit' the whole series of deductions of Part I, but only revisit the axioms (II §5), and a sample deduction.

Comma categories of the special kind E/X were introduced by the Grothendieck group, who used them for converting generalized elements into global elements, as in II §6 here. The density notion employed in II §7 is Isbell's 'adequacy' notion, and is equivalent to Ulmer's density notion.

Geometric theories, and their special role in topos theory, were recognized by Joyal, Reyes, .., cf. [56] p. 71 (where they are called 'coherent' theories).

Further historical information and bibliography may be found in [22], notably Chapter V.

III (general). All the models considered here are examples of 'gros toposes'. Gros toposes were discovered and utilized by the group around Grothendieck in the early 60's. A gros topos is a category of set valued functors on a small category, R, where R^{op} is a category of objects of geometric nature; R^{op} could be the category of smooth manifolds, say. If R is the category of finitely presented k-algebras, R^{op} is also a geometric-natured category, namely that of affine schemes over k. This category $\underline{Set}^R$ and some closely related categories were utilized by the group around Cartier and Grothendieck as the 'world' in which algebraic geometry lives, cf. notably [10], [9], and [78]. Also, Lawvere's first work on categorical dynamics was inspired by [10] and [9], cf. [50].

However, except for [78], the main emphasis in algebraic geometry has been on another 'world', namely that of schemes, which are certain topological spaces equipped with a sheaf of local rings. Likewise, Dubuc [12] advocated the consideration of C^∞-schemes (certain topological spaces equipped with a sheaf of Π_∞-algebras).

Synthetic differential geometry lives in these scheme-contexts, too, but we have consistently avoided them in favour of the gros toposes, because the latter are conceptually simpler, and the methods of categorical logic (cf. Part II) work better there. Anyway, the two approaches are well related, cf. [9] and [12].

Note that the category $\underline{\text{Set}}^{\Delta^{op}}$ of CSS-complexes is also a kind of gros topos, with very primitive model objects: affine simplices, forming the category Δ.

III 1. Theorem III 1.2 is from [38]. The Corollary that $R^D \simeq R \times R$ in $\underline{\text{Set}}^{\mathcal{R}}$ is much older, cf. [50] and [31].

III 2. The notion of ε-stability was found by me in 1977 (cf.[33]) for purposes not directly related to differential geometry, but shortly after, I found Theorem III 2.5. The deeper Theorem III 2.6 was then conjectured by me, and proved by Coste, Coste-Roy and me, and by Dubuc and Reyes, [6], [14].

III 3.-4. Well adapted models were first constructed by Dubuc [11] (in fact, the specific model $\mathcal{C}_C$ is from that paper). Essential for this was the replacing of the algebraic theory of $\mathbb{R}$-algebras by the richer algebraic theory $\mathbb{T}_\infty$; this theory was advocated already in Lawvere's 1967 lecture. Using $\mathbb{T}_\infty$, Dubuc could combine the (dual of the) category of manifolds, and the category of Weil-algebras into the category $\mathcal{C}_C$. The content of §§ 3.-4. if from [36], which in turn contains a streamlining of some of the results and notions of [11], as well as the Comparison Theorem (Theorem 3.2).

III 5. The results here are due to Dubuc, or are classical.

III 6. The notion 'ideal of local character' (= germ determined ideal) is due to Dubuc, [12], so far I know. The results of the paragraph are also due to him, except Theorem 6.6 which was proved by Lawvere (1980); Theorem 6.3 is a slight strenthening of Dubuc's [12], Proposition 12.

<u>III 7</u>. The open cover topology on $\mathcal{B}$ is described by Dubuc in [13], where Theorem 7.4 is also proved. The open cover topology appears in different form also in his [11] and [12].

<u>III 8</u>. Theorem 8.1 and 8.4 are due to Dubuc [12]; the proof we give for validity of Axiom B is somewhat different from his.

<u>III 9</u>. The notion of W-determined ideal was advocated by Reyes [68]. Joyal conjectured, and Reyes proved, [69], that the W-determined ideals are precisely the closed ideals for the "Whitney-" (Frechet-space) topology on $C^{\infty}(\mathbb{R}^n)$. Theorem 9.4 is a variation of Dubuc's results [11]. Theorem 9.5 and 9.6 are due to Porta and Reyes [65].

<u>III 10</u>. The conclusion of Theorem 10.1 was proved for the generic local ring by me in [30]. The proof presented for $R \in \mathcal{E}$ here is virtually the same. Theorem 10.3 is due to Penon [64], and is expected to lead to a very interesting development.

<u>III 11</u>. The results here are due to Kock, Porta, and Reyes [44], [65].

APPENDIX A: FUNCTORIAL SEMANTICS

We consider functorial semantics on three "levels", (or in three "doctrines"): algebraic theories, left-exact categories, and proper-left-exact categories.

Recall [49] that a (finitary) algebraic theory $\mathbb{T}$ is a category whose objects are 0,1,2,..., and where n is an n-fold product of 1 with itself n times, by means of specified $proj_i : n \to 1$ ($i=1,\dots,n$). The examples that occur in this book are $\mathbb{T}_k$ (for k a commutative ring in Set), and $\mathbb{T}_\infty$.

The theory $\mathbb{T}_k$ has for its morphisms $n \to m$ m-tuples of polynomials over k in n variables $X_1,\dots X_n$; they need not all occur. The composition is formal substitution of polynomials into polynomials. The $proj_i : n \to 1$ is just the polynomial X_i.

The theory $\mathbb{T}_\infty$ has for its morphisms $n \to m$ the set of smooth ($=C^\infty$) maps $\mathbb{R}^n \to \mathbb{R}^m$, with evident composition. So $\mathbb{T}_\infty$ is a full subcategory of Mf, the category of manifolds and smooth maps.

Note that $\mathbb{T}_k$ was syntactically defined, whereas $\mathbb{T}_\infty$ was semantically defined. Note also that there are natural inclusions

$$\mathbb{T}_{\mathbb{Z}} \subseteq \mathbb{T}_{\mathbb{Q}} \subseteq \mathbb{T}_{\mathbb{R}} \subseteq \mathbb{T}_\infty,$$

the last because a polynomial over any infinite field k is completely determined by the polynomial map $k^n \to k$ which it defines.

If $\mathbb{T}$ is an algebraic theory and $\mathcal{E}$ is a category with finite products, the category of $\mathbb{T}$-algebras in $\mathcal{E}$, denoted $\mathbb{T}\text{-Alg}(\mathcal{E})$, is the category of finite-product-preserving functors $R : \mathbb{T} \to \mathcal{E}$. The object $R(1) \in \mathcal{E}$ is the underlying object of the algebra R, and often is itself denoted R.

For instance, if $\mathbb{T} = \mathbb{T}_{\mathbb{Z}}$, the category of $\mathbb{T}$-algebras in E is equivalent to the category of commutative ring objects in E; if R is a ring object in E, we get a functor $\bar{R}: \mathbb{T} \to E$ by sending $\varphi \in \mathbb{T}(n,1)$ into the map $R^n \to R$ with description

$$(x_1, \ldots, x_n) \mapsto \varphi(x_1, \ldots, x_n)$$

which makes sense because φ is a polynomium in n variables.

If E = Sets and $\mathbb{T}$ is any algebraic theory, the functor

$$\hom_{\mathbb{T}}(n,-): \mathbb{T} \to \underline{\text{Sets}}$$

is a $\mathbb{T}$-algebra, denoted F(n). It solves the universal problem justifying the name "free algebra in n generators". The n generators are $\text{proj}_1, \ldots, \text{proj}_n$. If $\mathbb{T} = \mathbb{T}_k$, $F(n) = k[X_1, \ldots, X_n]$.

It is known that the category $\mathbb{T}$-Alg(Set) is complete and cocomplete. A $\mathbb{T}$-algebra B in Set is called finitely presented if there exists a coequalizer in $\mathbb{T}$-Alg(Set)

$$F(m) \rightrightarrows F(n) \to B$$

(a "finite presentation of B"). There is a weaker notion: B is said to be finitely generated if there exists a surjective $\mathbb{T}$-algebra homomorphism $F(n) \to B$. The full subcategories of $\mathbb{T}$-Alg(Set), consisting of finitely presented (respectively finitely generated) algebras, are denoted $FP\mathbb{T}$ and $FG\mathbb{T}$, respectively. There is a full and faithful functor

$$F: \mathbb{T} \to (FP\mathbb{T})^{op}$$

given by $n \mapsto F(n)$. It preserves finite products. It has an important universal property. First we note that the subcategory $FP\mathbb{T} \subseteq \mathbb{T}$-Alg(Set) is closed under finite colimits; in particular, $FP\mathbb{T}$ has finite colimits, so $(FP\mathbb{T})^{op}$ has finite inverse limits.

Let us call a category which has finite inverse limits <u>left exact</u>, and let us call a functor which preserves finite inverse limits <u>left exact</u>. Then

Theorem A.1. Given a left exact category E. Then, to any $\mathbb{T}$-algebra $R:\mathbb{T}\to E$, there exists a left exact functor $\bar{R}:\mathrm{FP}\mathbb{T}^{\mathrm{op}}\to E$ such that the diagram

$$
\begin{array}{ccc}
\mathbb{T} & \xrightarrow{\;\;R\;\;} & E \\
{\scriptstyle F}\searrow & \nearrow{\scriptstyle \bar{R}} & \\
(\mathrm{FP}\mathbb{T})^{\mathrm{op}} & &
\end{array}
$$

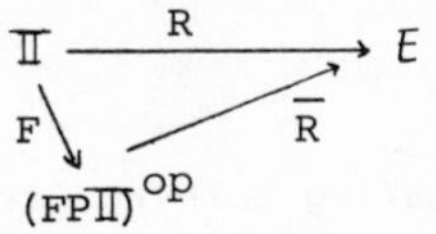

commutes. It is unique up to unique isomorphism, in fact, we have the stronger statement: given any functor $S:(\mathrm{FP}\mathbb{T})^{\mathrm{op}}\to E$ and any natural transformation

$$\tau:S\circ F \Rightarrow R = \bar{R}\circ F,$$

there exists a unique natural transformation $\bar{\tau}:S\Rightarrow\bar{R}$ with $\bar{\tau}_{F(n)}=\tau_n \;\forall n=0,1,2,\ldots$. Also, if $\nu:S\Rightarrow\bar{R}$ is a natural transformation with $\nu_{F(1)}=\tau_1$, then $\nu=\bar{\tau}$.

This Theorem is implicitly in [18], and (for the existence of $\bar{R}$) explicitly in [41]. The unique existence of $\bar{\tau}$ given τ is straight forward, and the uniqueness of $\bar{\tau}$ just knowing τ_1 follows because $R(n)=R(1)\times\ldots\times R(1)$.

If R is a $\mathbb{T}$-algebra in a left exact category E, one sometimes writes $\mathrm{Spec}_R(B)$ for $\bar{R}(B)$ $(B\in\mathrm{FP}\mathbb{T})$.

A Corollary of the Theorem is that

$$\mathbb{T}\text{-Alg}(E) \simeq \mathrm{Lex}((\mathrm{FP}\mathbb{T})^{\mathrm{op}},E),$$

the category of left exact functors $(\mathrm{FP}\mathbb{T})^{\mathrm{op}}\to E$. One may express this by saying that $(\mathrm{FP}\mathbb{T})^{\mathrm{op}}$ is the "theory of $\mathbb{T}$-algebras in the

doctrine of left exactness".[18] is the book about this doctrine, and also about the doctrine of proper-left-exactness:

A monic map $g:A \to B$ in a category is called *proper* monic if for any commutative square

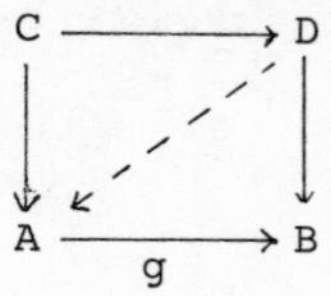

with the top map epic, there is a unique $D \to A$ making both resulting triangles commute. A *proper subobject* of an object B is a subobject represented by a proper monic. A category $\mathcal{A}$ is called *proper-left-exact* if it is left exact and if every class of proper subobjects of an object B has a limit; such limit is then necessarily a proper subobject of B and will be called a *proper* intersection. Similarly, we define the notion of *proper-left-exact* functor: a left exact functor preserving proper intersections.

The notions of proper epic, proper quotient, proper co-intersection, and proper-right-exact category/functor are defined dually.

The proper epics in $\mathbb{T}$-Alg(*Set*) are exactly those homomorphisms which are surjective on the underlying-set level. The same holds for $FG\mathbb{T}$.

The category $\mathbb{T}$-Alg(*Set*) has proper cointersections, and

$$FG\mathbb{T} \subseteq \mathbb{T}\text{-Alg}(\underline{Set})$$

is closed under formation of these, and also under the formation of finite colimits.

The functor $\mathbb{T} \xrightarrow{F} FP\mathbb{T}^{op} \to FG\mathbb{T}^{op}$ does for the doctrine of proper-left-exactness what F does for the doctrine of left exactness:

Theorem A.2. Replace $FP\mathbb{T}$ in Theorem A.1 by $FG\mathbb{T}$, and the two occurrences of the word 'left exact' by 'proper-left-exact'.

Then the Theorem still holds.

This is also implicitly in [18]. Again, one sometimes writes $\mathrm{Spec}_R(B)$ for $\overline{R}(B)$ $(B \in FG\Pi)$. If $B \in FP\Pi$, this is consistent with the previous usage.

Let E be cartesian closed and left exact. Let $R \in \Pi\text{-Alg}(E)$. We write R-Alg for $R{\downarrow}\Pi\text{-Alg}(E)$, and $\underset{\sim\sim}{\mathrm{Hom}}_{R\text{-Alg}}$ for the corresponding internal-hom-object formation, cf. I §12, II §4, and III §1. For any such R, any $C \in R\text{-Alg}$ and $B \in FP\Pi$, there is a map $\nu_{B,C}$ in E

$$\underset{\sim\sim}{\mathrm{Hom}}_{R\text{-Alg}}(R^{\overline{R}(B)}, C) \xrightarrow{\nu_{B,C}} \overline{C}(B) \tag{A.1}$$

where $\overline{R}$ and $\overline{C}$ are the functors $FP\Pi^{op} \to E$ associated to R and C by Theorem A1. To describe $\nu_{B,C}$ for all $B \in FP\Pi$, it suffices, by Theorem A1, to describe $\nu_{F(n),C}$ for all $n \in \Pi$, and prove naturality in n. We now describe $\nu_{F(n),C}$. We substitute F(n) for B, so $\overline{R}(B) = R^n$, $\overline{C}(B) = C^n$. So we should describe

$$\mathrm{Hom}_{R\text{-Alg}}(R^{R^n}, C) \xrightarrow{\nu_{F(n),C}} C^n$$

We do it for the case $E = \underline{\mathrm{Set}}$. (By the technique of Part II, this actually gives a description for any E). The i'th component of $\nu_{F(n),C}$ is the map

$$g \longmapsto g(\mathrm{proj}_i)$$

where g denotes an element of $\underset{\sim\sim}{\mathrm{Hom}}_{R\text{-Alg}}(R^{R^n}, C)$. The naturality w.r.t. $n \in \Pi$ follows because g is a Π-homomorphism.

Finally, if E is also proper-left exact, we may similarly construct $\nu_{B,C}$ for any $B \in FG\Pi$.

APPENDIX B: GROTHENDIECK TOPOLOGIES

Let $\mathcal{E}$ be a category. To equip $\mathcal{E}$ with a Grothendieck <u>pre-topology</u> means to give, for each $B \in \mathcal{E}$, a class $\mathrm{Cov}(B)$ of "covering families of B" or just "coverings of B"; the elements of $\mathrm{Cov}(B)$ must be families of arrows

$$\{\xi_i : X_i \to B \mid i \in I\} \tag{B.1}$$

with B as their common codomain. The data of the covering families should satisfy the following:

(1) If (B.1) is a covering of B, and $f: A \to B$ is an arbitrary arrow, then for each $i \in I$, the pull-back of ξ_i along f exists in $\mathcal{E}$,

$$\begin{array}{ccc} A \times_B X_i & \longrightarrow & X_i \\ {\scriptstyle f^*(\xi_i)}\downarrow & & \downarrow{\scriptstyle \xi_i} \\ A & \xrightarrow[f]{} & B \end{array}$$

and $\{f^*(\xi_i) \mid i \in I\}$ is a covering of A. ("Stability of the covering notion under pull-back".)

(2) If (B.1) is a covering of B, and, for each $i \in I$

$$\{\eta_{ij} : Y_j \longrightarrow X_i \mid j \in J_i\}$$

is a covering of X_i, then the family

$$\{Y_j \xrightarrow{\eta_{ij}} X_i \xrightarrow{\xi_i} B \mid i \in I, \ j \in J_i\}$$

is a covering of B (stability under composition).

(3) The one-element family $\{id_B : B \to B\}$ is a covering of B.

Grothendieck pretopologies give rise to Grothendieck topologies, a more general and invariant notion, which we shall not need. Grothendieck topologies may be described without any postulate about existence of sufficiently many pull-backs. The notion of sheaf for a Grothendieck topology j may be described in terms of any pretopology giving rise to j.

Definition. Let E be a category equipped with a Grothendieck pretopology. A functor $F: E^{op} \to \underline{Set}$ is a sheaf for it if any $B \in E$ and any covering (B.1) of it, the diagram

$$F(B) \longrightarrow \prod_{i \in I} F(X_i) \rightrightarrows \prod_{(i,j) \in I \times I} F(X_i \times_B X_j) \tag{B.2}$$

is an equalizer (where the three maps in an evident way are induced using the contravariant functorality of F).

The full subcategory of $\underline{Set}^{E^{op}}$ consisting of the sheaves is denoted $\tilde{E}$. A Grothendieck topos is a category of form $\tilde{E}$, where E is some small category equipped with a Grothendieck pretopology. The inclusion functor $\tilde{E} \hookrightarrow \underline{Set}^{E^{op}}$ has (for E small) a left adjoint a (a reflection functor), which preserves finite $\varprojlim$ (and arbitrary colimits, of course). This implies that E has very good exactness properties, much like the category $\underline{Set}$. Also, E is stably cartesian closed.

If the Yoneda embedding $E \to \underline{Set}^{E^{op}}$ factors through $\tilde{E}$, or equivalently, if each representable functor

$$\hom_{\mathcal{E}}(-,C) : \mathcal{E}^{op} \to \underline{\text{Set}}$$

for $C \in \mathcal{E}$, is a sheaf, one says that the Grothendieck pretopology is <u>subcanonical</u>. Note that monicness of the first map in (B.2) implies that a covering family for a subcanonical topology is necessarily jointly epic.

APPENDIX C: CARTESIAN CLOSED CATEGORIES

Let E be a category with finite products. We say E is cartesian closed if for $D \in E$, the functor $-\times D\colon E \to E$ has a right adjoint (denoted $(-)^D$). So there is a bijection (for all $X \in E$, $R \in E$):

$$\hom_E(X, R^D) \overset{\lambda_{X,R}}{\cong} \hom(X \times D, R),$$

natural in X and R, called "λ-conversion" or "exponential adjointness".

The category Set of sets is an example; here, R^D is the set of maps from D to R.

Any topos is cartesian closed. If E is a topos and $X \in E$, E/X is known to be a topos, in particular cartesian closed. It follows that any topos E is stably cartesian closed: E/X is cartesian closed $\forall X \in E$.

REFERENCES

[1] M. Artin, A. Grothendieck, and J.L. Verdier, Théorie des topos et cohomologie etales des schémas (SGA 4), Vol.1, Lecture Notes in Math. 269, Springer Verlag 1972.

[2] J. Bénabou, Catégories et logiques faibles, Oberwolfach Tagungsbericht 30/1973.

[3] M. Bunge, Models of differential algebra in the context of synthetic differential geometry, Cahiers de Top. et Géom. Diff. 22 (1981), 31-44.

[4] K.-T. Chen, Iterated path integrals, Bull. Amer. Math. Soc. 83 (1977), 831-879.

[5] M. Coste, Logique d'ordre superieur dans les topos élémentaires, Seminaire Benabou, Paris 1974.

[6] M. Coste and M.-F. Coste(-Roy), The generic model of an ε-stable geometric extension of the theory of rings is of line type, in Topos Theoretic Methods in Geometry, Aarhus Math. Inst., Var.Publ. Series No. 30 (1979).

[7] M. Coste and G. Michon, Petit et gros topos en géométrie algébrique, Cahiers de Top. et Géom. Diff. 22 (1981), 25-30.

[8] H.S.M. Coxeter, The real projective plane, (2nd ed.), Cambridge University Press 1961.

[9] M. Demazure and P. Gabriel, Groupes algébriques, North Holland Publishing Co. 1970.

[10] M. Demazure and A. Grothendieck, Schémas en Groupes, (SGA 3) I, Lecture Notes in Mathematics 151, Springer Verlag 1970.

[11] E.J. Dubuc, Sur les modèles de la géometrie différentielle synthétique, Cahiers de Top. et Géom. Diff. 20 (1979), 231-279.

[12] E.J. Dubuc, C^{∞}-schemes, American Journ. Math. (to appear).

[13] E.J. Dubuc, Open covers and infinitary operations in C^{∞}-rings, Aarhus Preprint Series 1980/81 No. 4, to appear in Cahiers de Top. et Géom. Diff.

[14] E.J. Dubuc and G.E. Reyes, Subtoposes of the ring classifier, in Topos Theoretic Methods in Geometry, Aarhus Math. Inst. Var.Publ.Series No. 30 (1979).

[15] C. Ehresmann, Categories in differential geometry, Cahiers de Top. et Géom. Diff. 14 (1973), 175-177.

[16] J. Emsalem, Géometrie des points épais, Bull.Soc.Math.France 106 (1978), 399-416.

[17] H. Federer, Geometric measure theory, Grundlehren der Math. Wiss. 153, Springer Verlag 1969.

[18] P. Gabriel and F. Ulmer, Lokal präsentierbare Kategorien, Lecture Notes in Mathematics 221, Springer Verlag 1971.

[19] C. Godbillon, Géométrie differentielle et mécanique analytique, Hermann, Paris 1969.

[20] M. Golubitsky and V. Guillemin, Stable mappings and their singularities, Graduate Texts in Mathematics 14, Springer Verlag 1973.

[21] J. Gray, The meeting of the Midwest Category Seminar in Zürich 1970, in Reports of the Midwest Category Seminar V, Lecture Notes in Mathematics 195, Springer Verlag 1971.

[22] J. Gray, Fragments of the history of sheaf theory, in Applications of Sheaves, Proceedings Durham 1977, Lecture Notes in Mathematics 753, Springer Verlag 1979.

[23] A. Grothendieck and J. Dieudonné, Eléments de géometrie algebrique, Publ. Math. IHES, 4, 8,11,... .

[24] V. Guillemin and A. Pollack, Differential topology, Prentice-Hall Inc. 1974.

[25] M. Hakim, Topos anneles et schemas relatifs, Ergebnisse der Math. 64, Springer Verlag 1972.

[26] T. Hjelmslev, Die natürliche Geometrie, Abh. Math. Sem. Hamburg 2 (1923), 1-36.

[27] J. Isbell, Adequate subcategories, Illinois J.Math. 4 (1960), 541-552.

[28] P. Johnstone, Topos Theory, London Math. Society Monographs 10, Academic Press 1977.

[29] F. Klein, Vorlesungen über höhere Geometrie, Grundlehren der math. Wiss. 22, Springer Verlag 1926.

[30] A. Kock, Universal projective geometry via topos theory, J. Pure Appl. Alg. 9 (1976), 1-24.

[31] A. Kock, A simple axiomatics for differentiation, Math. Scand. 40 (1977), 183-193.

[32] A. Kock, Taylor Series calculus for ring objects of line type, Journ. Pure Appl. Alg. 12 (1978), 271-293.

[33] A. Kock, Formally real local rings, and infinitesimal stability, in Topos Theoretic Methods in Geometry, Aarhus Math. Inst. Var. Publ. Series No. 30 (1979).

[34] A. Kock, On the synthetic theory of vector fields, in Topos Theoretic Methods in Geometry, Aarhus Math. Inst. Var. Publ. Series No. 30 (1979).

[35] A. Kock, Formal manifolds and synthetic theory of jet bundles, Cahiers de Top. et Géom. Diff. 21 (1980), 227-246.

[36] A. Kock, Properties of well adapted models for synthetic differential geometry, Journ. Pure Appl. Alg. 20 (1981), 55-70.

[37] A. Kock, Differential forms with values in groups, Aarhus Preprint Series 1980/81 No. 5. Preliminary report in Cahiers de Top. et Géom. Diff.

[38] A. Kock, A general algebra-geometry duality, Lecture at 17. Peripatetic Seminar on Sheaves and Logic, Sussex 1980. To appear in Proc. 19. Peripatetic Seminar on Sheaves and Logic, Paris-Nord 1981.

[39] A. Kock (ed.), Topos theoretic methods in geometry, Aarhus Math.Inst. Var.Publ. Series No. 30 (1979).

[40] A. Kock, P. Lecouturier and C.J. Mikkelsen, Some topos theoretic concepts of finiteness, in Model Theory and Topoi, Lecture Notes in Math 445, Springer Verlag 1975.

[41] A. Kock and G.E. Reyes, Doctrines in categorical logic, in Handbook of Mathematic Logic, North Holland Publishing Company 1977.

[42] A. Kock and G.E. Reyes, Manifolds in formal differential geometry, in Applications of Sheaves, Proceedings Durham 1977, Lecture Notes in Math. 753, Springer Verlag 1979.

[43] A. Kock and G.E. Reyes, Connections in formal differential geometry, in Topos theoretic Methods in Geometry, Aarhus Math. Inst. Var. Publ. Series No. 30 (1979).

[44] A. Kock and G.E. Reyes, Models for synthetic integration theory, Math.Scand. 48 (1981).

[45] A. Kock, G.E. Reyes, and B. Veit, Forms and integration in synthetic differential geometry, Aarhus Preprint Series 1979/80 No. 31.

[46] A. Kumpera and D. Spencer, Lie equations, Annals of Mathematics Studies 73, Princeton 1972.

[47] S. Lang, Algebra, Addison-Wesley Publishing Co. 1965.

[48] F.W. Lawvere, An elementary theory of the category of sets, Proc.Nat.Acad.Sci. USA 52 (1964), 1506-1511.

[49] F.W. Lawvere, Some algebraic problems in the context of functorial semantics of algebraic theories, in Reports of the Midwest Category Seminar II, Lecture Notes in Mathematics 61, Springer Verlag 1968.

[50] F.W. Lawvere, Categorical Dynamics, in Topos Theoretic Methods in Geometry, Aarhus Math.Inst. Var.Publ. Series No. 30 (1979).

[51] F.W. Lawvere, Toward the description in a smooth topos of the dynamically possible motions and deformations of a continuous body, Cahiers de Top. et Géom.Diff. 21 (1980), 377-392.

[52] S. Lie, Allgemeine Theorie der partiellen Differentialgleichungen erster Ordnung, Math.Ann. 9 (1876), 245-296.

[53] S. Lie, Vorlesungen über Differentialgleichungen, Leipzig 1891.

[54] S. Lie, Geometrie der Berührungstransformationen, Leipzig 1896.

[55] S. MacLane, Categories for the working mathematician, Graduate Texts in Mathematics No. 5, Springer Verlag 1971.

[56] M. Makkai and G.E. Reyes, First Order Categorical Logic, Lecture Notes in Mathematics, 611, Springer Verlag 1977.

[57] B. Malgrange, Ideals of differentiable functions, Tata Institute Bombay/Oxford University Press 1966.

[58] B. Malgrange, Equations de Lie, I, Journ.Diff.Geom. 6 (1972), 503-522.

[59] W. Mitchell, Boolean topoi and the theory of sets, J.Pure Appl. Algebra 2 (1972), 261-274.

[60] J. Munkres, Elementary differential topology, Annals of Mathematics Studies 54, Princeton University Press 1963.

[61] E. Nelson, Tensor Analysis, Mathematical Notes, Princeton University Press 1967.

[62] Ngo Van Que and G.E. Reyes, Théorie de distributions et théorème d'extension de Whitney, Exposé 8, in Geom. Diff. Synthétique, Fasc. 2, Rapport de Recherches 80-12, U. de Montréal D.M.S. 1980.

[63] G. Osius, A note on Kripke-Joyal semantics for the internal language of topoi, in Model Theory and Topoi, Lecture Notes in Mathematics 445, Springer Verlag 1975.

[64] J. Penon, Infinitesimaux et intuitionnisme, Cahiers de Top. et Géom. Diff. 22 (1981), 67-72.

[65] H. Porta and G.E. Reyes, Variétés à bord et topos lisses, Exposé 7 in Geom. Diff. synthétique, Fasc. 2, Rapport de Recherches 80-12, U. de Montréal D.M.S. 1980.

[66] G.E. Reyes, Cramer's rule in the Zariski topos, in Applications of sheaves, Proceedings, Durham 1977, Lecture Notes in Math. 753, Springer Verlag 1979.

[67] G.E. Reyes, Théorie des modèles et faisceaux, Advances in Math. 30 (1978), 156-170.

[68] G.E. Reyes, Analyse C^∞, Exposé 4, in Geom.Diff.Synthétique, Fasc. 2, Rapport de Recherches 80-12, U. de Montréal D.M.S. 1980.

[69] G.E. Reyes, Analyse dans les topos lisses, Cahiers de Top. et Géom.Diff. (to appear).

[70] G.E. Reyes (ed.), Géometrie Différentielles Synthétique, Rapport de Recherches du Dépt. de Math. et de Stat. 80-11 and 80-12, Université de Montréal, 1980.

[71] G.E. Reyes and G.C. Wraith, A note on tangent bundles in a category with a ring object, Math.Scand. 42 (1978), 53-63.

[72] G. de Rham, Variétés différentiables, Hermann, Paris 1960.

[73] A. Robinson, Non standard analysis, North-Holland Publ.Co. 1966.

[74] M. Spivak, Differential Geometry (Second Edition), Publish or Perish, Berkeley 1979.

[75] J.C. Tougeron, Idéaux de fonctions différentiables, Ergebnisse der Math. 71, Springer Verlag 1972.

[76] F. Ulmer, Properties of dense and relative adjoint functors, J. Algebra 8 (1967), 77-95.

[77] B.L. van der Waerden, Algebra, Die Grundlehren der math. Wissenschaften 33, Springer Verlag 1960.

[78] W.C. Waterhouse, Introduction to Affine Group Schemes, Graduate Texts in Mathematics 66, Springer Verlag 1979.

[79] A. Weil, Théorie des point proches sur les variétés différentiables, in Colloq. Top. et Géom.Diff., Strassbourg 1953.

[80] H. Whitney, Differentiable even functions, Duke J.Math. 10 (1943), 159-160.

[81] G.C. Wraith, Generic Galois theory of local rings, in Applications of Sheaves, Durham 1977, Lecture Notes in Math. 753, Springer Verlag 1979.

INDEX